SpringerBriefs in Applied Sciences and Technology

PoliMI SpringerBriefs

Series Editors

Barbara Pernici, Milan, Italy
Stefano Della Torre, Milan, Italy
Bianca M. Colosimo, Milan, Italy
Tiziano Faravelli, Milan, Italy
Roberto Paolucci, Milan, Italy
Silvia Piardi, Milan, Italy

For further volumes:
http://www.springer.com/series/11159

Francesco Amigoni · Viola Schiaffonati
Editors

Methods and Experimental Techniques in Computer Engineering

POLITECNICO
DI MILANO

Editors
Francesco Amigoni
Viola Schiaffonati
Dipartimento di Elettronica
  Informazione e Bioingegneria
Politecnico di Milano
Milan

Italy

ISSN 2282-2577   ISSN 2282-2585   (electronic)
ISBN 978-3-319-00271-2   ISBN 978-3-319-00272-9   (eBook)
DOI 10.1007/978-3-319-00272-9
Springer Cham Heidelberg New York Dordrecht London

Library of Congress Control Number: 2013941941

Printed on acid-free paper

Springer is part of Springer Science+Business Media (www.springer.com)

Preface

Computing and science reveal today a synergic relationship, in particular when considering the nature and role of methods and techniques adopted in computer science and engineering. Is computing a science? Does computing adopt the scientific method? Can computing help science? Is science a computational process? To what extent is science supported by computing? These are just a few of the many questions we can ask when recognizing the synergic relationship between computing and science.

Computing plays an increasing role in the scientific endeavor, both in modeling and simulating entities and in practically supporting various scientific activities. Think for example of autonomous artificial systems playing an increasing role in the discovery and verification of scientific hypotheses (robot scientists); or of computer and robotic systems used to discover and test mechanistic hypotheses about intelligent and adaptive behaviors of living systems; or of computational tools applied to chemical and toxicological discovery process; or of computational models used in earth system sciences to represent, understand, predict, and manage the behavior of environmental systems. Moreover, the role of scientific method in computing is getting increasingly important, especially in providing ways to experimentally evaluate the properties of complex computing systems, such as in drawing inspiration from scientific experiments to assess and benchmark autonomous systems. Indeed, these artificial systems are so huge and intricate, and can have highly unpredictable interactions with the external environment that even their designers do not know in advance their behavior in every situation.

This volume presents these issues from a conceptual and methodological perspective by addressing specific case studies at the intersection between computing and science. More precisely, we intend computing both as an infra-science and as a science, and we organize the contributions accordingly. In the first role, computing is a tool for scientific disciplines (chemistry, biology, environmental sciences, animal behavior, ...) and the privileged direction of the knowledge flow is from computing to science. In the second role, computing is a science, even with some important peculiarities, and the privileged direction of the knowledge flow is from science (in particular the experimental scientific method) to computing. The volume attempts to offer an insight into how computing and science can influence each other both at a practical level and at a theoretical one. Since we cannot cover all the issues related to the common ground between computing and science, we go in-depth for some significant questions, on the basis of the specific case studies presented.

This volume stems from the experience of the course "Computing and Science: What Can They Do for each Other?" for graduate students in the Ph.D. program in Information Engineering held at Politecnico di Milano during the Spring of 2012. As for the course, our aim here is to present an integrated overview over the relationships between computing and science that goes beyond their specific field of study.

Part I of the volume investigates the impact of computing on different scientific fields. Chapter 1 illustrates, by means of examples taken from various environmental contexts, how numerical computing has unlocked new uses for existing environmental models with terrific impacts from both the scientific and engineering perspectives. Indeed, ever growing computing power and its increasing availability are revolutionizing the way environmental models are constructed and used. Besides model use, this chapter considers how numerical computing has also significantly impacted model construction, especially in the context of the so-called empirical models, which simply could not exist without the computer programs used for their construction. Chapter 2 describes the basis of computational chemistry and how computational methods have been applied to biology, and to toxicology in particular. It discusses how and why the experimental practice in biological science is moving toward computational modeling and simulation to confirm hypotheses, to provide data for regulation, and to help in designing new chemicals.

Part II of the volume analyzes the possible role of scientific concepts, like the experimental scientific method, on some fields of computing, in particular on autonomous robotics, which shows challenging issues related to the interaction between computational entities, the autonomous robots, and the real world. In this part, by investigating to what extent traditional experimental principles can be applied to some fields of computer science and engineering, we argue in favor of the importance of a methodological characterization of computing, and of computer engineering in particular, considering the role played by experiments. This is also a way to reflect on the status of the discipline not from the perspective of its object, but from the perspective of its method, whereas the debate on the disciplinary status of computer science has been mostly revolved around the issues on what computer science is (science, engineering, or else) and whether computing is a scientific discipline. Chapter 3 presents an analysis of experimental trends in autonomous robotics by surveying the autonomous robotic articles presented over the last dozen of years at the International Conference on Autonomous Agents and Multiagent Systems (AAMAS). By starting from some experimental methodologies proposed for autonomous robotics, this chapter reflects on the way they are implemented in the current robotic research practice. Chapter 4 takes a further step in this direction by illustrating the RAWSEEDS toolkit for benchmarking robots that perform simultaneous localization and mapping activities. The chapter shows both the benefits and the difficulties of building tools for the rigorous experimental evaluation of robotic performance.

Part III of the volume presents an attempt to investigate on the relationship between computing and science while considering, at the same time, computational tools as support to experimentation and some experimental principles as inspiration for the development of good experimental methodologies in computing. Chapter 5 presents some interesting roles played by biorobotics in the study of intelligent and adaptive behavior, where biorobotic experiments can give rise to different theoretical outcomes, such as the evaluation of the plausibility of a hypothesis or the formulation of new scientific questions. Moreover, this chapter illustrates some methodological and epistemological problems raised by biorobotics, arguing that dealing with these problems is crucial to justify the idea according to which robotic implementation and experimentation can offer interesting theoretical contributions to the study of intelligence and cognition.

Finally, some concluding remarks recall that one of the goals of this volume is to leave the reader with the recognition that many of the issues at the intersection between computing and science are still open and represent excellent topics for future investigations.

We would like to thank all the contributors, and the speakers and students of the Ph.D. course from which this work originated. We are especially grateful to Mark Bedau for his support to this initiative.

Milan, July 2013

Francesco Amigoni
Viola Schiaffonati

Contents

Part I From Computing to Science

1 **Computational Models for Environmental Systems** 3
   Francesca Pianosi

2 **How Far Chemistry and Toxicology are Computational
   Sciences?** 15
   Giuseppina Gini

Part II From Science to Computing

3 **Good Experimental Methodologies for Autonomous Robotics:
   From Theory to Practice** 37
   Francesco Amigoni, Viola Schiaffonati and Mario Verdicchio

4 **RAWSEEDS: Building a Benchmarking Toolkit for Autonomous
   Robotics** 55
   Giulio Fontana, Matteo Matteucci and Domenico G. Sorrenti

Part III Computing and Science: Back and Forth

5 **Biorobotics: A Methodological Primer** 71
   Edoardo Datteri

Concluding Remarks 87

Contributors

Francesco Amigoni Dipartimento di Elettronica, Informazione e Bioingegneria, Politecnico di Milano, Piazza Leonardo da Vinci 32, 20133 Milan, Italy, e-mail: francesco.amigoni@polimi.it

Edoardo Datteri Dipartimento di Scienze Umane per la Formazione "R. Massa", Università degli Studi di Milano-Bicocca, Piazza dell'Ateneo Nuovo 1, 20126 Milan, Italy, e-mail: edoardo.datteri@unimib.it

Giulio Fontana Dipartimento di Elettronica, Informazione e Bioingegneria, Politecnico di Milano, Piazza Leonardo da Vinci 32, 20133 Milan, Italy, e-mail: giulio.fontana@polimi.it

Giuseppina Gini Dipartimento di Elettronica, Informazione e Bioingegneria, Politecnico di Milano, Piazza Leonardo da Vinci 32, 20133 Milan, Italy, e-mail: giuseppina.gini@polimi.it

Matteo Matteucci Dipartimento di Elettronica, Informazione e Bioingegneria, Politecnico di Milano, Piazza Leonardo da Vinci 32, 20133 Milan, Italy, e-mail: matteo.matteucci@polimi.it

Francesca Pianosi Dipartimento di Elettronica, Informazione e Bioingegneria, Politecnico di Milano, Piazza Leonardo da Vinci 32, 20133 Milan, Italy; Department of Civil Engineering, University of Bristol, University Walk, Bristol BS8 1TR, UK, e-mail: francesca.pianosi@polimi.it; francesca.pianosi@bristol.ac.uk

Viola Schiaffonati Dipartimento di Elettronica, Informazione e Bioingegneria, Politecnico di Milano, Piazza Leonardo da Vinci 32, 20133 Milan, Italy, e-mail: viola.schiaffonati@polimi.it

Domenico G. Sorrenti Dipartimento di Informatica, Sistemistica e Comunicazione, Università degli Studi di Milano-Bicocca, Piazza dell'Ateneo Nuovo 1, 20126 Milan, Italy, e-mail: domenico.sorrenti@unimib.it

Mario Verdicchio Dipartimento di Ingegneria, Università degli Studi di Bergamo, via Salvecchio 19, 24129 Bergamo, Italy, e-mail: mario.verdicchio@unibg.it

Part I
From Computing to Science

Computational tools seem today essential to the conducting of science. Even without considering e-science applications and scenarios, computing plays an ever growing role at the service of most scientific fields. Part I of this volume focuses on the role of computing as infra-science, emphasizing its instrumental role both to support and to enlarge the current scientific practice. The cases discussed in the two chapters composing this part move from environmental science, showing significant examples of how computational tools are used to model and analyze environmental systems, to chemistry and toxicology, showing how these sciences are massively heading toward computer-based models and simulations. Both the chapters forming Part I aim not only at discussing relevant and successful application cases of computer-supported science, but also at reflecting on the methodological issues—still largely open—raised by the adoption of these computational tools.

Chapter 1
Computational Models for Environmental Systems

Francesca Pianosi

Abstract Mathematical models are increasingly used to represent, understand, predict and manage environmental systems. In recent years, increasing data availability and computing power have induced dramatic changes in the way environmental models are developed and used. Models are being constructed at increasingly high resolution and complexity, while computer-based simulations allow for unprecedented uses of environmental models and data. Nonetheless, the complexity of environmental systems and the uncertainty in environmental data still pose a number of challenges in the construction, validation and use of environmental models. In this chapter, the relation between observations, modelling and numerical computing, and their implications in the environmental domain will be discussed through several examples. Without pretending to be exhaustive, the selection of examples mainly aim at highlighting the variety of contexts, application domains and modelling purposes affected by new computing technology.

Keywords Environmental modelling · Physically-based and data-driven modelling · Environmental uncertainty

1.1 Introduction

Mathematical models are increasingly used to represent, understand, predict and manage environmental systems, including the atmospheric system, water resources, ecosystems, agro-forestry systems, etc. [1]. The application domains and purposes of environmental models are manifold [2]. In some contexts, models are used to test

F. Pianosi (✉)
Dipartimento di Elettronica, Informazione e Bioingegneria, Politecnico di Milano, Milan, Italy
e-mail: francesca.pianosi@polimi.it

Department of Civil Engineering, University of Bristol, Bristol, UK
e-mail: francesca.pianosi@bristol.ac.uk

F. Amigoni and V. Schiaffonati (eds.), *Methods and Experimental Techniques in Computer Engineering*, PoliMI SpringerBriefs,
DOI: 10.1007/978-3-319-00272-9_1, © The Author(s) 2014

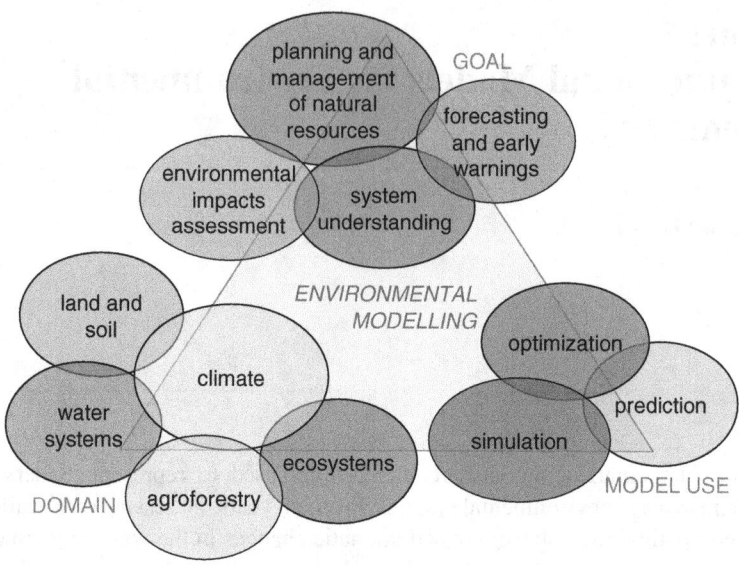

Fig. 1.1 Application domain, use and goal of environmental models

hypothesis, validate theories and ultimately improve our understanding of natural processes. Alternatively, models may be used for forecasting future system conditions or for assessing their response to human interventions via model simulation and what-if analysis. In many applications, models are expected to inform and support decision-making processes, from early warning systems to the sustainable planning and management of natural resources (Fig. 1.1).

Ever growing computing power on the one hand, and increasing data availability on the other, are revolutionizing the way environmental models are constructed and used. Advances in pervasive sensor networks [3] and remote sensing techniques [4] can provide environmental data at local and global scale at increasingly higher temporal and spatial resolution. Cheaper and more easily accessible computing facilities allow for processing this huge data flow and including them into more and more sophisticated models and innovative applications [5]. In this chapter the contribution of numerical computing to environmental modelling will be discussed through several examples. Without pretending to be exhaustive, the selection of examples mainly aim at highlighting the variety of contexts, application domains and modelling purposes affected by new computing technology.

The example in Sect. 1.2 comes from the field of ecology. The use of mathematical models to describe population dynamics has a long history and started much before the advent of computers.[1] The equations appearing in these models can sometimes

---

[1] The first quantitative population growth model can be traced back to the *Essay on the Principle of Population* by Thomas Robert Malthus in 1798. Another classical topic of ecology is the prey-predator dynamics, that was first modeled by Alfred J. Lotka and Vito Volterra in the 1920s.

be solved by pencil and paper, at least under specific conditions, for instance at equilibrium. However, computer-based numerical integration has allowed for solving ecological models virtually at any conditions, and this in turn has dramatically enlarged our possibility of testing models and their underlying theories, as shown in this example.

The example in Sect. 1.3 deals with water systems. Here again the mathematical models used to describe rivers and reservoirs translate simple hydraulic principles that have been known for decades, while the reservoir operation model is based on the Bellman optimality principle [6] dating back to the late 1950s. However, the implementation of these simulation and optimization models at the resolution scale used in this example would have been impossible up to few years ago because of limited computing power.

These examples illustrate how numerical computing has unlocked new uses of existing models with dramatic impacts from both the scientific and engineering perspective. Besides model use, numerical computing has also significantly impacted model construction. This is true in many different instances but is especially remarkable in the context of the so called *empirical models*, which simply could not exist without the computer programs that are used for their construction. This is discussed in Sect. 1.4.

Still, the improvement in our modelling capacity made possible by increasing availability of data and computing power is likely to confront limits in the environmental domain. The complexity of the investigated processes and the uncertainty in the data are so high that computers are not seemingly to resolve all the conflicting issues inherent in the construction and use of environmental models. This is discussed in Sect. 1.5.

1.2 Numerical Computing and New Understanding of Environmental Systems: Scientific Perspective

The first example comes from the domain of quantitative ecology. Ecological models aim at providing quantitative explanations and predictions of the relationships among living organisms and the environment, including infra-specific and interspecific processes like population growth, stability and extinction, competition for available resources, predation, symbiosis, parasitism, etc. One issue that has puzzled scientists in the domain of aquatic ecology is the so called "paradox of the plankton". The paradox consists in a mismatch between model predictions and real-world observations. Competition models predict that, at equilibrium, the number of coexisting species cannot exceed the number of the limiting resources they consume. However, for phytoplankton species the number of limiting resources (e.g., nitrogen, phosphorus, silicon, iron) is very low while empirical observations show that dozens of phytoplankton species coexist. A very interesting contribution to the debate comes from [7], that provides a solution to the plankton paradox made possible by computer-based simulations.

The competition model is a set of differential equations that describe the rate of growth of the n phytoplankton species and of the k resources. Equations are derived by translating basic principles of mortality, fertility and competition into mathematical relations. By setting to zero all the differential equations, one can derive the equilibrium state of the system. This is the condition when population and resource densities are constant. Usually it is assumed that the ecosystem should spontaneously tend towards this condition and return to it when recovering from external disturbances. Given the relative simplicity of the model, the equilibrium solution can be derived analytically with pencil and paper. It says that the populations with nonzero density (i.e., that can coexist at equilibrium) are in number lower or equal to the number k of resources. The conclusion has been named the "principle of competitive exclusion" and, as anticipated, it is in contrast with empirical experience.

Alternative models have been proposed to explain the observed species diversity of planktonic communities. They include factors external to the phytoplankton dynamics, like selective predators, spatial heterogeneity, or time-varying weather conditions. By using computer-based simulation, instead, [7] demonstrates that the paradox can be resolved using the original competition model, provided that one looks at its simulated behaviour when the equilibrium is not reached. Specifically, numerical integration of the model equations show that: first, when there are at least three species and three resources, the equilibrium may not be reached and the simulated species densities may exhibit oscillations; second, the number of species that keep oscillating in time and do not go to extinction may be higher than the number of resources. As oscillations are not generated by other sources of external variability, this study demonstrates that competition theory is in fact sufficient to explain the coexistence of phytoplankton species.

1.3 Numerical Computing and New Understanding of Environmental Systems: Engineering Perspective

The second example comes from the domain of integrated water resources management. Mathematical models are widely used in this sector to support the operation and planning of water resources systems like rivers, lakes, wetlands, and human artifacts like dams, irrigation and drainage systems, urban supply networks, etc.

The example here reported is an application developed by the author and others, to the efficient use of regulated lakes and reservoirs. Reservoirs may enhance the economic, social and environmental value of watersheds by enabling water reallocation in space and time. However, often watersheds comprise multiple reservoirs that are operated independently one to another to meet different targets. The lack of coordination generates inefficiency and economic loss and induces conflicts among different water uses.

The lake Como watershed considered in [8] is a typical example. The water system develops along the river Adda in Northern Italy, with a topology common to many Alpine watersheds: a large storage capacity distributed in many hydropower reservoirs in the upper watershed region; a large regulated lake in the middle region; and multiple water consumption users, mostly farmers, in the lower region. Spring snowmelt is the most important contribution to the creation of the seasonal storage, which is reallocated over time according to two different strategies. The lake regulation exploits the accumulated volume in the summer to supply downstream irrigation, while hydropower operators keep their storages full up until the following winter when the demand for energy peaks and the production is more valuable. This results in the potential for conflict between farmers and hydropower companies, which is highest in particularly dry summers, when farmer associations claim that water shortages could be mitigated if the water retained by the hydropower companies were available.

In the study [8], mathematical models are used to compare the system performances under different institutional settings and thus assess the space for improving the overall system efficiency. First, the study simulates the hypothetical condition where a single super-operator has full access to the system data conditions and makes all the decisions simultaneously, balancing upstream (hydropower) and downstream (irrigation) interests. Simulation experiments show that under such a centralized approach there exists a win-win solution in which the irrigation deficit can be significantly reduced without economic loss in the hydropower production. In other terms, the study demonstrates that the main limitations to the current system performance do not stem from physical constraints (e.g., limited storing capacity) but they come from the institutional, legal and operational framework. Subsequently, the centralized operating policy is analyzed in order to gain insights into suitable strategies to foster cooperation among the involved agents. The analysis suggests a coordination mechanism based on constraining the minimum release of upstream hydropower reservoirs in particularly critical situations. Simulations show that this coordinated approach, although suboptimal with respect to the ideal centralized solution, still can significantly improve the historical operation.

The study combines simulation models of the physical components like reservoirs, hydropower plants and the river network, and optimization models that can mimic the decision-making process of reservoir operators. The application of the latter is particularly challenging from the computational standpoint. The optimization algorithm there used, stochastic dynamic programming, has computational complexity that grows exponentially with the number of system variables, so that the application to a multi-reservoir network like the one in [8] would have been computationally unaffordable only few years ago.

Fig. 1.2 Classification of
models in the environmental
domain

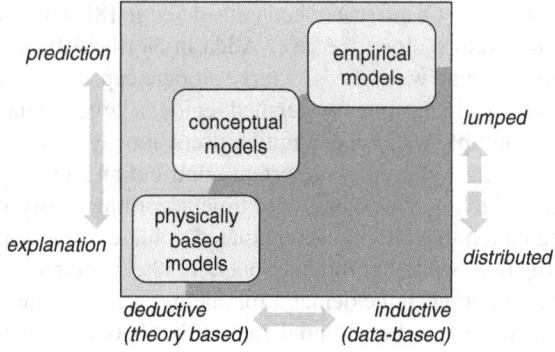

1.4 Numerical Computing and New Modelling Paradigms

Environmental models are often divided into three categories, *physically-based*, *conceptual* and *empirical*. The classification is based on the way models are constructed, i.e., by deduction from a scientific theory or by induction from the data, but also reflects the purpose they are constructed for, i.e., explanation or prediction (Fig. 1.2). Numerical computing has had a significant and sometimes predominant influence on the development and use of all the three model categories.

Physically-based models provide a detailed description of the processes occurring in the system. They consider both temporal and spatial variability and thus take the form of a set of partial differential equations with distributed parameters. Model identification is based on the translation of physical principles into the model equations, and parameter values are mainly decided based on field measurements or laboratory experiments. Historically, the progress in the development of such models was limited by our imperfect knowledge of environmental processes as well as by computer capability. Cheaper and more easily accessible computing power has allowed for simultaneously increasing the number of processes reproduced in the model, their spatial resolution and the length of simulation horizons. For instance, [9] reviews the advances of physically-based climate models boosted by the advent of supercomputers.

Conceptual models give a simplified description of the system functioning. They reproduce only the main processes and usually neglect the spatial dimension or give it a very simplified representation. The model variables represent spatial averages of the quantities of interest and their temporal dynamics is given by a set of differential equations with lumped parameters. Since parameters are not associated to measurable quantities their value must be guessed or inferred from data by minimizing the distance between model outputs and observations. The data-based parameter estimation exercise, or model calibration, can be carried out manually or by means of automatic procedures. Ever increasing availability of data and computing power has dramatically encouraged the use of automatic calibration procedures and nowadays most of the software packages implementing environmental models also include specific routines for automatic calibration.

As increasing computing power has made model calibration techniques faster and faster, data-based inference can be used not only to derive parameter estimates but also the model structure itself, by repetition and comparison of calibration results under different hypothesized model structures. The models so obtained are called *empirical*, or data-driven or black-box. The latter term highlights the purely predictive nature of these models, which relate system inputs and outputs without any attempt at reproducing the inner system processes. The use of empirical models in the environmental domain has often been questioned because they are deemed to be hardly understood and trusted by their expected users. Nonetheless, they have gained more and more attention because they are usually cheaper to develop and easier to use, requiring less data and computing power at both stages, while providing comparatively good performances at least for operational purposes. Note that for empirical models, numerical computing is not simply a factor influencing the accuracy, usability or range of application of the model, but rather it is the essential reason for their very existence.

Information and communications technology is likely to further shape the way models are constructed, used and verified in even more complex and unpredictable fashions. Mobile devices and online applications can be used to give people an active role in the collection and verification of data that are lately used to feed models, and to disseminate and verify the information derived by models. For instance, [10] reviews a number of recent applications in the water domain, where mobile phones are used for gathering user-recorded water level data, for providing model-based advices to farmers, for disseminating flood forecasts, etc. Social computing will allow for new ways of eliciting distributed knowledge and expertise that can be used to verify and complement with analytic knowledge embodied in mathematical models [11]. Although these applications are still at a prototype stage, their development is gaining increasing interest from researchers and entrepreneurs, and in the next future they may challenge the traditional boundaries between expert-based and mathematical models, as well as between physically-based, conceptual and empirical models.

1.5 Environmental Models, Data and Uncertainty

The discussion in the previous section seems to suggest that increasing computing power can only lead to expand our modelling capability. However, this may be not so obvious, due to the high level of system complexity and uncertainty in the data. In environmental systems many different physical, biological, chemical processes overlap and influence each other, variables are strongly heterogenous in time and space, controlled experiments are often impossible, observations are highly uncertain, even the very definition of the system boundaries is critical and sometimes arbitrary since interfaces between climate, water, soils, vegetation, etc. are not clearcut and reciprocal influences may be negligible or not depending on the scale of interest. Finally,

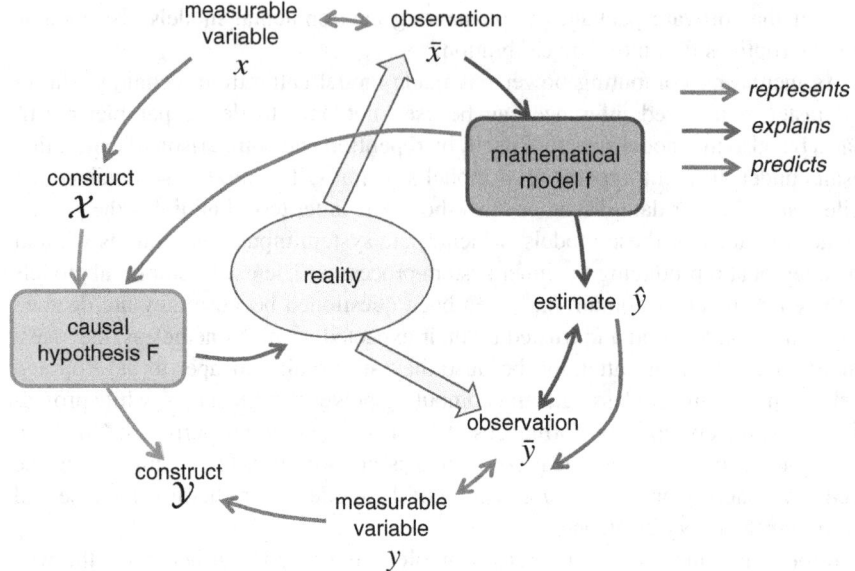

Fig. 1.3 Processes, variables and observations in the construction of models and theories

the influence of human behavior on the earth systems has become so broad and deep in the last decades that scientists have coined the term *Anthropocene* to refer to the geological era we are living in [12], and consequently, in many instances, natural processes cannot be investigated independently from socio-economic ones [13], which adds up new dimensions of complexity to the problem of modelling environmental systems.

In such a context, the fit-to-data, which is the guiding principle of model validation, can be critical. To discuss this topic, consider the formalization in Fig. 1.3, which is a personal elaboration of the one given in [14]. A mathematical model f is expected to provide a relation between an input variable x and an output variable y. In *predictive modelling*, model f is used to produce output predictions. In *explanatory modelling*, model f is used to test the causal hypothesis F. In other words, f is obtained by translation of F into mathematical equations and tested against data, so that if model f fits the data, then the underlying theory F is deemed to explain reality.

Because of measurement errors, the measured input $\bar{x}$ that feeds the model differs from the real input x, and the model estimate $\hat{y}$ must be evaluated against the observed output $\bar{y}$ instead of the actual output y. Still, if measurement errors are small, x and $\bar{x}$, or y and $\hat{y}$ can be confounded, as we usually do for instance when we say that we predict a *variable* while we actually predict its *measurement*. Also, in formulating the causal hypothesis F, inputs and outputs are usually given in terms of "theoretical constructs" $\mathcal{X}$ and $\mathcal{Y}$ (see again [14] and reference herein), which later are associated to some measurable variables x and y. For instance, in hydrology, theoretical constructs may be the rainfall over a catchment (input) and the river flow (output), while measurable variables are rainfall and flow at some specific sites where climate and hydrological stations are located. So again there is a slight shifting in meaning when we say that we use models to reproduce "natural processes" because actually we use models to reproduce variables (y) that we consider representative of processes $(\mathscr{Y})$. And in practice we end up in representing data $(\bar{y})$ instead of variables, because data is the only thing we possess.

The confusion is acceptable when the choice of observable variables is robust and observation errors are small, so that observations $\bar{y}$ and $\bar{x}$ are really representative of processes $\mathscr{X}$ and $\mathscr{Y}$. However, this is often not the case in environmental modelling, where system complexity makes the choice of model variables far from obvious, and measurement errors are usually large and sometimes even in the same order of magnitude as the measured variable.[2]

In such a context, can we conclude that, say, the predictive model f_1 is preferable over the predictive model f_2, based on the fact that the predictions $\hat{y}_1$ by f_1 are closer to observations $\bar{y}$, when measurement errors $(y - \bar{y})$ may be the same order of magnitude as the difference between model predictions $(\hat{y}_1 - \hat{y}_2)$? Or, should we reject theory F because the associated model f does not fit well the data, when we know that those data contain large errors? And maybe the very choice of the model variables x and y as representative of constructs $\mathscr{X}$ and $\mathscr{Y}$ is doubtful?

In practice, such controversies are resolved by environmental modellers based on their judgement and personal experience. Often, models f are trusted because the underlying theories F. In other words, the explanatory process is reversed and, instead of looking at the model's fit to data in order to confirm the theory, it is the theory that is used to motivate the credibility of the model, even against observations. In fact, a low fit-to-data of model f is often taken as an evidence of poor data quality rather than of a weak underlying theory. Sometimes this is justified based on previous successful applications of the theory to other case studies. Still, the issue about which should be trusted more, theory or data, is open and underlies the long-standing controversy between supporters and opponents of the empirical modelling approach [18].

According to Beven [19], new computing technology will contribute to unravel the matter. As more and more data and computing power become available and virtually every place can be represented by mathematical models, emphasis will

---

[2] Uncertainty stems from multiple sources including inaccuracy of measuring devices, errors introduced by preprocessing, problems of sampling. The quantification of such uncertainty has become an important research field. Some examples from the water systems domain are given in [15–17]. Mittelbach et al. [17] provides an example of problems arising in the calibration of measuring devices. They investigate the accuracy of soil moisture sensors and show that in field conditions none of the evaluated sensors has a level of performance consistent with the respective manufacturer specifications developed in laboratory conditions. Baldassarre and Montanari [16] investigates preprocessing errors and define a methodology for quantifying the uncertainty of river flow data, which are obtained from river stage data by application of rating curves. They show that in their case study, overall errors may be as large as 42.8 % (at the 95 % confidence level) with an average value of 25.6 %. An example of sampling difficulties is given in [15] when discussing river quality modelling, where monitoring programs have sampling frequency and location designed for regulation purposes and often inadequate to capture the system dynamics as required for successful model identification.

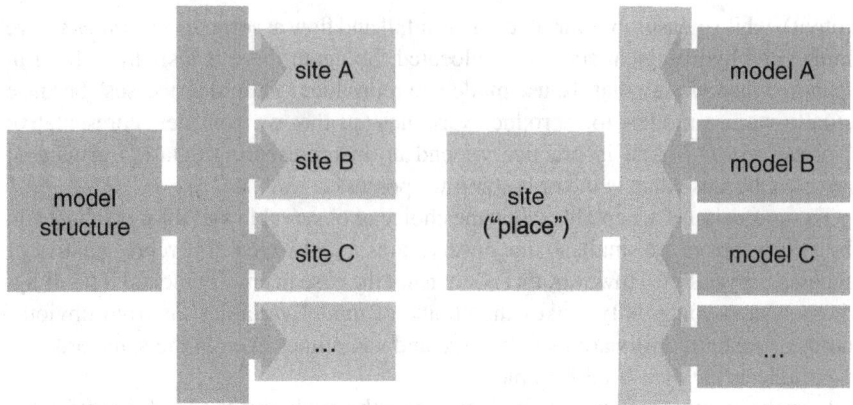

Fig. 1.4 From learning about processes to learning about places

change from a process of learning about model structures (theories) to a process of learning about the specific features of particular places. In other words, if in the past environmental modellers used to search for generalized model structures that could be efficiently adjusted (calibrated) and applied to multiple sites, exchanging accuracy for generality, in the next future the abundance of data and computing power will make it possible to investigate each specific site "from scratch", testing different model structures and designing the ad hoc combination that guarantees the best possible performance for a specific site and a specific purpose (Fig. 1.4). In the words of [19], "a new generation of environmental models [will appear] that are geared towards the management of specific places rather than general process representation".

1.6 Conclusions

Computer technology is changing the way environmental models are constructed and used in many respects. Existing models are being used in new ways to perform simulation/optimization experiments in a wider range of conditions or at more realistic resolution steps, providing new insights on system functioning and thus contributing to advance our scientific knowledge as well as supporting better management of natural resources. New models are being built with increasing complexity of structures while empirical models are becoming a viable alternative to traditional physically-based models. Still, because of the complexity of the investigated systems, the multiplicity of interacting components, and the high level of uncertainty in environmental data, the identification and validation of environmental models can be regarded as a "wicked problem" that rarely has a true-or-false solution. The choice of the most appropriate model structure or parameterization is not univocal and it may vary with the scale and purpose of the modelling exercise and change over time as new knowledge becomes available. Human expertise is likely to keep playing a crucial role in the construction of environmental models and the interpretation of their results. By multiplying the mechanisms for information gathering and dissemination, information and communications technology itself will possibly contribute to reinforcing the integration of human and computer intelligence in the environmental modelling domain.

References

1. Wainwright J, Mulligan M (2004) Environmental modelling: finding simplicity in complexity. Wiley, Chichester
2. Jakeman A, Lecther R, Norton P (2006) Ten iterative steps in development and evaluation of environmental models. Environ Model Softw 21:602–614
3. Hart J, Martinez K (2006) Environmental sensor networks: a revolution in the earth system science? Earth Sci Rev 78(3–4):177–191
4. Butler D (2007) Earth monitoring: the planetary panopticon. Nature 450:778–781
5. Lux T, Sydow A (2005) Special issue on environmental modelling. ERCIM News 61
6. Bellman R (1959) Dynamic programming. Princeton University Press, Princeton
7. Huisman J, Weissing F (1999) Biodiversity of plankton by species oscillations and chaos. Nature 402(25):407–410
8. Anghileri D, Castelletti A, Pianosi F, Soncini-Sessa R, Weber E (2013) Optimizing watershed management by coordinated operation of storing facilities. J Water Res (Pl-ASCE In press), doi: 10.1061/(ASCE)WR.1943-5452.0000313
9. Washington W, Buja L, Craig A (2008) The computational future for climate and earth system models: on the path to petaflop and beyond. Philos Trans R Soc A 367:833–846
10. Jonoski A, Alfonso L, Almoradie A, Popescu I, van Andel S, Vojinovic Z (2012) Mobile phone applications in the water domain. Environ Eng Manag J 11(5):919–930
11. Fraternali P, Castelletti A, Soncini-Sessa R, Ruiz CV, Rizzoli A (2012) Putting humans in the loop: social computing for water resources management. Environ Model Softw 37:68–77
12. Crutzen PJ, Stoermer EF (2000) Global change newsletter. The Anthropocene 41:17–18
13. Editorial (2011) Welcome to the anthropocene. The Economist May 28
14. Shmueli G (2010) To explain or to predict? Stat Sci 25(3):289–310
15. Berthouex P, Brown L (1994) Statistics for environmental engineers. CRC Press, Boca Raton
16. Di Baldassarre G, Montanari A (2009) Uncertainty in river discharge observations: a quantitative analysis. Hydrol Earth Syst Sci 13:913–921
17. Mittelbach H, Lehner I, Seneviratne S (2012) Comparison of four soil moisture sensor types under field conditions in Switzerland. J Hydrol 430–431:39–49
18. Young P, Parkinson S, Lees M (1996) Simplicity out of complexity in environmental modelling: Occam's razor revisited. J Appl Stat 23(2–3):165–210
19. Beven K (2007) Towards integrated environmental models of everywhere: uncertainty, data and modelling as a learning process. Hydrol Earth Syst Sci 11(1):460–467

Chapter 2
How Far Chemistry and Toxicology are Computational Sciences?

Giuseppina Gini

Abstract In this chapter we describe the basis of computational chemistry and discuss how computational methods have been extended to biology, and toxicology in particular. Since about 20 years, chemical experimentation is more and more replaced by modelling and virtual experimentation. Computer modelling of biological properties is still a debated topic. However, the need of safety assessment of chemicals is pushing toxicology towards computer modelling. The term *in silico* discovery is now applied to chemical design, to computational toxicology, and to drug discovery. We discuss how the experimental practice in biological science is moving more and more towards computer modelling and simulation. Such virtual experiments confirm hypotheses, provide data for regulation, and help in designing new chemicals.

Keywords *In silico* experiments · Chemoinformatics · QSAR · Toxicology

2.1 Introduction

"All science is computer science". When a New York Times article in 2001 used this title, the general public was aware that the introduction of computers has changed the way that experimental sciences develop. A first example of the historical connection between chemistry and computer science is the development of fragment codes, usually called fingerprints, used to filter large data sets of molecules for the presence or absence of a particular sub-structure. For that M. Lynch, J. Ziv, and A. Lempel produced the Ziv-Lempel algorithms, which are the basis of the wide used algorithms for data compression [1].

Giuseppina Gini (✉)
Dipartimento di Elettronica, Informazione e Bioingegneria, Politecnico di Milano,
Milan, Italy
e-mail: giuseppina.gini@polimi.it

F. Amigoni and V. Schiaffonati (eds.), *Methods and Experimental Techniques in Computer Engineering*, PoliMI SpringerBriefs,
DOI: 10.1007/978-3-319-00272-9_2, © The Author(s) 2014

Chemistry and physics are among the best examples of such a new way of making science. A new discipline, *chemoinformatics* has been in existence for the past two decades [2, 3]. Many of the activities performed in chemoinformatics are information retrieval [4], aimed at searching for new molecules of interest when a single molecule has been identified as being relevant. However chemoinformatics is more than "chemical information"; it requires strong algorithmic development. Simulations now are widely used in chemistry, material sciences, engineering, and process control, to name a few fields. What about life sciences?

Starting from chemistry and reviewing the role of mathematics and then algorithms in chemical research, in this chapter we will move to some part of biological experimentation. In particular we will see how animal experiments, aimed at providing a standardized result about a biological property, as bioavailability or even death, can be mimicked by new *in silico* methods. Our emphasis is on toxicology and QSAR (Quantitative Structure Activity Relationships) methods [5–7].

The aim of this chapter is to briefly review how and why life sciences are moving more and more towards modelling and simulation. Here computing is a tool for scientific disciplines, with the interesting consideration that many basic computing tools (as graph representation, simulators, efficient data hashing) had their origin in the arising needs of reasoning with atoms and molecules.

In Sect. 2.2 we introduce the role of computer-based models and algorithms in chemistry. In Sects. 2.3 and 2.4 we see how biological modelling and toxicology has evolved from the first animal models to the *in silico* models. Section 2.5 presents the problems related to human toxicology and drug development. Section 2.6 discusses environmental toxicology and risk assessment. Section 2.7 points out the pitfalls and new trends of *in silico* methods, while Sect. 2.8 presents some conclusions.

2.2 Chemistry

The word 'chemistry' is from the Greek khymeia ($\chi\upsilon\mu\epsilon\iota\alpha$) "to fuse together". For centuries it has been seen as a kind of magic, a way to transform elements. All the activities connected to chemistry, from cooking to metallurgy, to medicinal remedia have been dominated by empirical rules. In his 1830 book *Course of Positive Philosophy* Auguste Compte wrote:

> Any attempt to use mathematical methods in the study of chemical problems should be considered not rationale and contrary to the spirit of chemistry. In case mathematical analysis would assume a prominent role in chemistry—an aberration that fortunately is almost impossible—that would bring to a rapid degeneration of this science.

2.2.1 Atoms and Elements

The basic components of matter have been investigated in the Greek philosophy. Democritos postulated the existence of atoms. "There are atoms and space.

Everything else is opinion". Aristotle declared the existence of four elements: fire, air, water, and earth, stating that all matter is made up of these four elements.

In much more recent times, atoms and elements have been scientifically defined. In 1789 Antoine Lavoisier published a list of 33 chemical elements grouped into gases, metals, non metals, earths. In 1803 John Dalton in his *Atomic Theory* stated that all matter is composed of atoms, which are small and indivisible.

Theoretically-based models of atoms were defined much later. Rutherford in 1909 adapted the solar system model: the atom is mostly empty space with a dense positively charged nucleus surrounded by negative electrons. In 1913 Bohr proposed that electrons travelled in circular orbits. Finally in the 1920's the electron cloud model was defined by Schrödinger; in it an atom consists of a dense nucleus composed of protons and neutrons surrounded by electrons. The most probable locations of the electron predicted by Schrödinger's equation coincide with the locations specified in Bohr's model.

About the elements, their accepted definition was the periodic table, published by Mendeleev in 1869. It organized the elements into a table, listing them in order of atomic weight, and starting a new row when the characteristics of the elements began to repeat [8]. It happened much before the development of theories of atomic structure; after that, it became apparent that Mendeleev had listed the elements in order of increasing atomic number.

So in the first 30 years of the XX century chemistry (and physics) has been refunded. Quantum theory and material physics have changed the way that materials are studied. Dirac, just one century after Compte, wrote:

> The physical rules necessary for a mathematical theory of the whole chemistry and of part of physics are known, and the only difficulty is that the application of those rules generates equations too complicated to be solved.

2.2.2 Computer-Based Representation for Molecules

A molecule is an electrically neutral group of atoms held together by covalent bonds. A molecule is represented as atoms joined by semi-rigid bonds.

The graph theory, established back in XVIII century, initially evolved through chemistry; the name 'graph' indeed derives from its use in drawing molecules. The valence model naturally transforms a molecule into a graph, where the atoms are represented as vertices and the bonds as edges. The edges are assigned weights according to the kind of bond. Today hydrogens are not represented in the graph since they are assumed to fill the unused valences [9]. This representation is called 2D chemical structure.

A common representation of the graph is the adjacency matrix, a square matrix with dimension N equal to the number of atoms. Each position (i, j) specifies the absence (0 value) or the presence of a bond connecting the atoms i and j, filled with 1, 2, 3 to indicate simple, double, or triple bond, 4 for amide bond, 5 for aromatic bond. An example of a matrix representation is in Fig. 2.1.

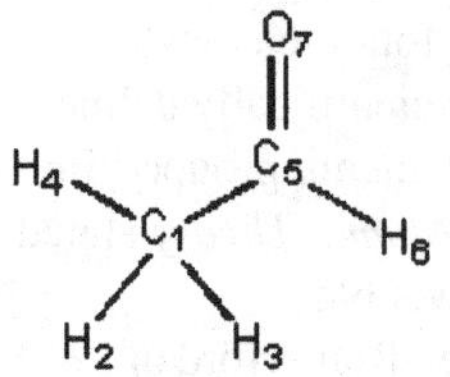

	C1	H2	H3	H4	C5	H6	O7
C1	0	1	1	1	1	0	0
H2	1	0	0	0	0	0	0
H3	1	0	0	0	0	0	0
H4	1	0	0	0	0	0	0
C5	1	0	0	0	0	1	2
H6	0	0	0	0	1	0	0
O7	0	0	0	0	2	0	0

Fig. 2.1 Graph representation and adjacency matrix of methil-aldehaide

To write the molecule as a text string, Simplified Molecular Input Line Entry Specification (SMILES) [10] is popular. SMILES is a context free language expressing the graph visit in a depth first style:

- Atoms are represented by their atomic symbols in upper-case.
- Bonds are Single, implicit; Double, "="; or Triple, "#".
- Branches are placed between round parentheses.
- Cycles are represented breaking one bond in each ring.

The SMILES notation suffers the lack of a unique representation, since a molecule can be encoded beginning anywhere. Therefore canonical SMILES [11] was proposed.

SMILES	NAME	FORMULA	GRAPH	3D
CC	Ethane	CH3CH3	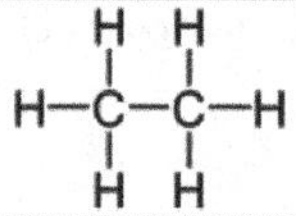	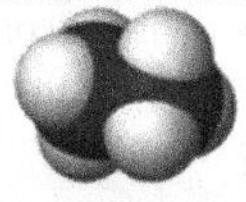
C=O O=C	Formaldehyde	CH2O	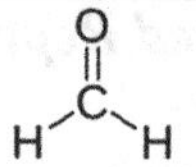	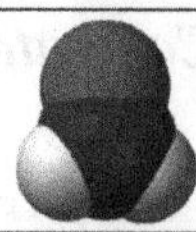
CCO OCC C(C)O C(O)C	Ethanol	CH3CH2OH	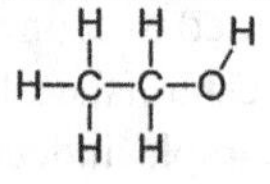	
c1ccccc1	Benzene	C6H6	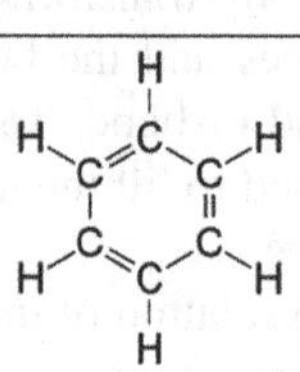	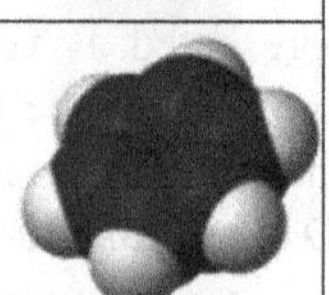

Fig. 2.2 Some 2D and 3D representations of molecules

(a)   **(b)**

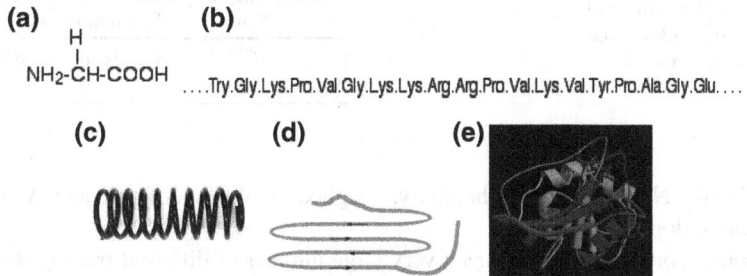

Fig. 2.3 a Glycine, an aminoacid. **b** A part of the primary structure of a protein. **c** The secondary structure alpha helixes. **d** The secondary structure beta sheets. **e** A tertiary structure from http://www.rcsb.org/pdb/

What about the real shape of molecules? They are 3D objects, and as such they should be represented. Figure 2.2 shows some examples of different representations. For biological molecules the representations should consider the real 3D shape in space. Figure 2.3 shows the aminoacid glycine, its primary structure (the sequence of aminoacids indicated by the first 3 letters of the name), the secondary structure, that is the organization of regions in alpha helixes and beta sheets, and 3D folding.

The point about defining the 3D shape of a molecule will take us to the basic methods of computational chemistry.

2.2.3 Computational Chemistry

Computational chemistry is a branch of chemistry that uses computers to assist in solving chemical problems, studying the electronic structure of matter. It uses the results of theoretical chemistry, incorporated into computer programs, to calculate chemical structures and properties. The methods cover both static and dynamic situations through accurate methods, *ab initio*, and less accurate methods, *semi-empirical*. *Ab initio* methods are based entirely on theory from first principles of quantum chemistry. Semi-empirical methods employ also experimental results from related molecules to approximate some parameters. *Ab initio* are computationally expensive, so the size of the molecules that can be modelled is limited to a few hundred atoms. For big molecules it is necessary to introduce empirical parameters. When the molecular system is even bigger, the simulation is statistically based.

Computational chemistry is a way to move away from the traditional approach of solving scientific problems using only direct experimentation, but it does not remove experimentation. Experiments produce new data; the role of theory is to situate all the new data into a framework, based on mathematical rules.

It is now possible to simulate with the computer an experiment before running it. It all happened in about 50 years, from the first *ab initio* calculations done in 1956

Table 2.1 The number of stars and molecules in real and virtual spaces

	Stars	Small molecules
Existent	10^{22}	10^7 currently in CAS registry
Virtual	0	10^{80}

at MIT to the Nobel prize for Chemistry, assigned in 1998 to J. Pople and W. Kohn for computational chemistry.

Through computer simulations a very large number of different but structurally related molecules are created using the methods of combinatorial chemistry, able to create a "library" of thousand of different but related compounds [12].

Today in the CAS (Chemical Abstracts Service) registry there are more than 71 millions of unique numerical identifiers assigned to every chemical described in the open scientific literature; the expected dimension of the chemical space is tens orders of magnitude bigger. Table 2.1 shows a comparison between the number of potential chemical compounds and the number of stars.

Advances in robotics have led to an industrial approach to combinatorial synthesis, enabling companies to routinely produce over 100,000 new and unique compounds per year. Today new products are designed and checked on the computer before they are synthesized in the laboratory. Those methods are part of the *in silico* methods that will be described in the following.

2.3 Biological Models and Toxicology

In antiquity, the physiology research was carried out on animals. Although these observations and their interpretations were frequently erroneous, they established a discipline. The explosion of molecular biology in the second half of the XX century increased the importance of in vivo models [13].

In biomedical research the investigations may be classified as observational or experimental.

- Observational studies are carried out when the variables influencing the outcomes of the phenomena under study cannot be controlled directly. These variables are observed and an attempt is made to determine the correlations between them.
- Experimental studies require to directly control selected variables and to measure the effects of these variables on some outcome. The results of experimental studies tend to be more robust compared to observational studies.

Experimental studies may be carried out on in vitro biological systems such as cells, microorganisms, tissue slice preparations. Experiments using in vitro systems are useful where the screening of large number of potential therapeutic candidates may be necessary, or in making fast tests for possible pollutants. In vitro systems are, however, non-physiological and have important limitations since their biological complexity is much lower than that of most of the animal species including humans. While data from experiments carried out in vitro can establish mechanisms, in vivo biological systems, using whole organisms, are required to study how such mechanisms behave in real conditions.

Models are meant to mimic the subject under study. Biomedical research models can also be either analogues or homologues.

- Analogous models relate one structure or process to another. An analogue animal model can explain some of the mechanisms of humans.
- Homologous models reflect the genetic sequence of the organism under study.

Often animal models are both analogues and homologues. Of course the ideal model for a human is a human, but animal models are so far the best approximation.

All models have their limitations; their prediction can be poor, and their transferability to the real phenomena they model can be unsatisfactory. So extrapolating data from animal models to the environment or to human health depends on the degree to which the animal model is an appropriate reflection of the condition under investigation. These limitations are, however, an intrinsic part of all modelling approaches. Most of the questions about animal models are ethical more than scientific. In public health the use of animal models is imposed by different regulations, and it is unlikely that any health authority will allow the adoption of novel drugs without supporting animal data.

2.3.1 Bioassays for Toxicity

Toxicity is the degree to which a substance can damage an organism. Paracelsus (1493–1541) wrote: "All things are poison and nothing is without poison; only the dose makes a thing not a poison." The relationship between dose and its effects on the exposed organism is of high significance in toxicology.

Animals have been used for assessing toxicity in pioneering experiments since more than one century. In more recent times the process of using animal testing to asses toxicity has been defined in the following way:

- Toxicity can be measured by its effects on the target.
- Because individuals have different levels of response to the same dose of a toxin, a population-level measure of toxicity is often used which relates the probabilities of an outcome for a given individual in a population. Example is median lethal dose LD_{50}: the dose that causes the death of 50% of the population after a specified test duration. Examples of some doses are in Table 2.2.
- When the dose is individuated, "safety factors" are defined. For example, if a dose is safe for a rat, one might assume that one tenth that dose would be safe for a human, allowing a safety factor of 10.

This process is based on assumptions that usually are very crude. It presents many open issues. For instance it is more difficult to determine the toxicity of chemical

Table 2.2 Examples of LD_{50} for common chemicals

Chemical	Target	LD_{50}
Water	Rat, oral	90000 mg/kg
Sucrose	Rat, oral	29700 mg/kg
Table salt $NaCl$	Rat, oral	3000 mg/kg
Paracetamol	Rat, oral	1944 mg/kg
Caffeine	Rat, oral	192 mg/kg
Nicotine	Rat, oral	50 mg/kg
Dioxin	Rat, oral	20 μg/kg

mixtures (gasoline, cigarette smoke, waste) since the percentages of the chemicals can vary, and the combination of the effects is not exactly a summation.

Perhaps the most common continuous measure of biological activity is the IC_{50} (inhibitory concentration), which measures the concentration of a compound necessary to induce a 50% inhibition of the biological activity under investigation.

The dose-response curve describes the change in effect on an organism caused by different levels of doses after a certain exposure time. A dose-response curve is a x-y graph relating the magnitude of a stressor to the response of the organism.

- The measured dose (usually milligrams per kilogram of body-weight for oral exposures) is plotted on the x axis and the response is plotted on the y axis.
- The response may be a physiological or biochemical response.
- LD_{50} is used in human toxicology; IC_{50}, inhibition concentration, and its dual EC_{50}, effect concentration, are fundamental to pharmacology.

Usually the logarithm of the dose is plotted on the x axis, so the curve is typically sigmoidal, with the steepest portion in the middle. In Fig. 2.4 we see an example of the dose response curve for LD_{50}.

Fig. 2.4 A curve for $\log(LD_{50})$

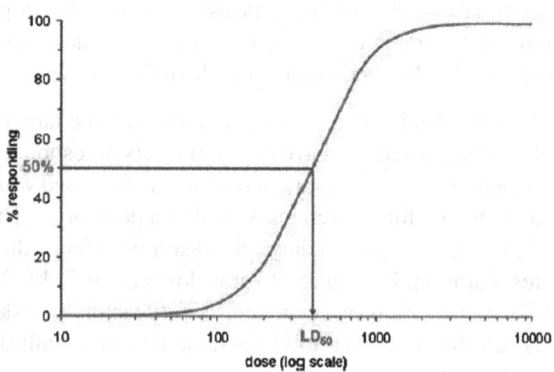

2.3.2 Animal Testing Versus in Vitro and in Silico Testing

Worldwide it is estimated that the number of vertebrate animals used annually in testing ranges from the tens of millions to more than 100 million. Animal testing was introduced for drugs, in particular in response to many deaths from the Sulfanilamide in 1937. In the 1960s, in reaction to the Thalidomide tragedy, further laws were passed requiring safety testing also on pregnant animals.

Those in vivo models give doses for some species, and are used to extrapolate data to human health or to the environment. As we said above, the extrapolation of data from species to species is not obvious. For instance, the lethal doses for rats and for mice are sometimes very different. In vitro toxicity testing is the scientific analysis of the effects of a chemical on cultured bacteria or mammalian cells. It is known that their results poorly correlate with the results of in vivo.

How to construct a model that relates a chemical structure to its effect was investigated even before computers were available. The term *in silico* covers the current methods devoted to this end.

2.4 *In Silico* Methods

The term '*in silico*' refers to the fact that computers are used, and computers have silicon in their hardware. The most known *in silico* methods are the QSAR (Quantitative Structure Activity Relationships) methods, derived from the suggestion made in 1868 by A. Crum Brown and T. Fraser that a mathematical relationship can be defined between the physiological action of a molecule and its chemical constitution [14].

2.4.1 QSAR

Given quantitative data, we can build a QSAR model that seeks to correlate our particular response with molecular descriptors that have been computed or even measured from the molecules themselves [15]. QSAR methods were first pioneered by Corwin Hansch in the 1940s, who analyzed congeneric series of compound and formulated the QSAR equation:

$$\log(1/C) = a \cdot \log P + b \cdot Hs + c \cdot Es + \text{const}$$

where C is effect concentration, $\log P$ is octanol-water partition coefficient, Hs is Hammett substituent constant (electronic), Es is Taft's substituent constant, and a, b, and c are parameters. The octanol-water partition coefficient $\log P$ is the ratio of concentrations of a compound in the two phases of a mixture of two immiscible solvents at equilibrium. It is a measure of the difference in solubility of the compound in these two solvents. With high octanol-water partition coefficient the chemical substance is hydrophobic and preferentially distributed to hydrophobic compartments such as cell membrane, while hydrophilic substances are found in hydrophilic compartments such as blood serum [16].

Sometimes the QSAR methods take more specific names as: QSPR (Quantitative Structure Property Relationship) or QSTR (Quantitative Structure Toxicity Relationship). They all correlate a dependent variable (the effect) with a set of independent variables (usually calculated properties, or descriptors).

2.4.1.1 Molecular Descriptors

The generation of informative data from molecular structures is of high importance in chemoinformatics. There are many possible approaches to calculating molecular descriptors [17], that represent local or global salient characteristics of the molecule. Different classes are:

- Constitutional descriptors, depending on the number and type of atoms, bonds, and functional groups.
- Geometrical descriptors, which give molecular surface area and volume, moments of inertia, shadow area projections, and gravitational indices.
- Topological indices, based on the topology of molecular graph [9]. Examples are the Wiener index (the sum of the number of bonds between all nodes) and the Randic index (the branching of a molecule).
- Physicochemical properties attempt to estimate the physical properties of molecules. Example are molecular weight, hydrogen bond acceptors, hydrogen bond donors, and partition coefficients, as $\log P$.
- Electrostatic descriptors, such as partial atomic charges, depending on the possibility to form hydrogen bonds.
- Quantum chemical descriptors, related to the molecular orbitals.
- Fingerprints. Since subgraph isomorphism (substructure searching) in large molecular databases is time consuming, substructure screening was developed as a rapid method of filtering out those molecules that definitely do not contain the substructure of interest. The method uses fingerprints, binary strings encoding a molecule, where the 1 or 0 in a position means whether the substructure of this position in the dictionary is present or not.

2.4.1.2 Model Construction

After selecting the relevant descriptors, whatever method is chosen to develop predictive models, it is important to take heed of the model quality statistics and ensure a correct modelling methodology is used, such as testing the model against an external and unseen test set to ensure it is not overfitting to the training set. Model extrapolation is another concern that frequently occurs when models are applied outside

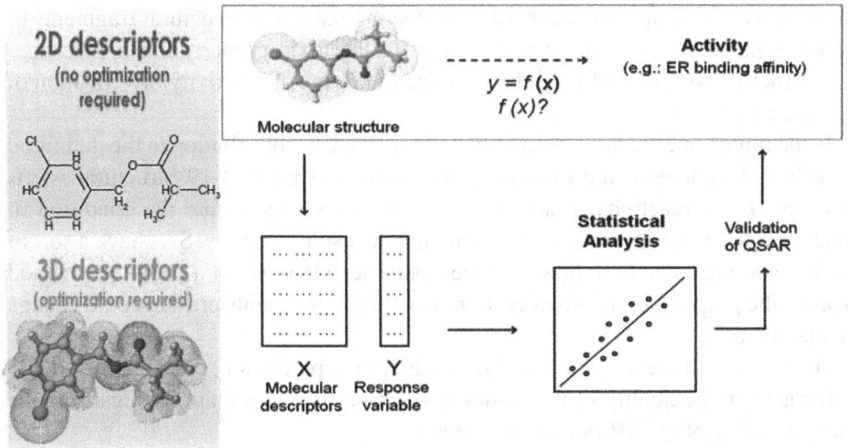

Fig. 2.5 The steps in constructing a QSAR model

the space from which the models were generated. Numerous model statistics are available that can indicate if new data points can be predicted by the model [18].

Two types of supervised learning methods are applied widely: classification and regression. Classification methods assign new objects, in our case molecules, to two or more classes—either biologically active or inactive. Regression methods attempt to use continuous data, such as a measured biological response variable, to correlate molecules with that data so as to predict a continuous numeric value for unseen molecules using the generated model [6]. Figure 2.5 illustrates the flow chart of the activities in the QSAR construction.

There is generally a trade-off between prediction quality and interpretation quality. Interpretable models are generally desired in situations where the model is expected to provide information about the problem domain. However, these models tend to suffer in terms of prediction quality as they become more interpretable. The reverse is true with predictive models, in that their interpretation suffers as they become more predictive. Models that are highly predictive tend to use molecular descriptors that are not readily interpretable by the chemist. However, predictive models are generally not intended to provide transparency, but predictions that are more reliable and can therefore be used as high-throughput models. If interpretability is of concern, other methods are available, more or less as a kind of expert systems, or SAR [19, 20].

2.4.2 SAR

SAR (Structure-Activity Relationships) typically makes use of rules created by experts to produce models that relate subgroups of the molecule atoms to a biological property. The SAR approach consists in detecting particular structural fragments of molecule already known to be responsible for the toxic property under investigation. Structural rules usually can be explained in terms of reactivity or activation of biological pathways.

In the mutagenicity/carcinogenicity domain, the key contribution in the definition of such toxicophores came from [21], who compiled a list of 19 Structural Alerts (SAs) for DNA reactivity. Practically SAs are rules which state the condition of mutagenicity by the presence of peculiar chemical substructures. SAs have a sort of mechanistic interpretation; however, their presence alone is not a definitive method to prove the property under investigation, since the substituents present could change the classification.

To conclude, if the main aim of (Q)SAR is simply prediction, the attention should be focused on the quality of the model, and not on its interpretation. Regarding the interpretability of QSAR models [14] states:

> The need for interpretability depends on the application, since a validated mathematical model relating a target property to chemical features may, in some cases, be all that is necessary, though it is obviously desirable to attempt some explanation of the mechanism in chemical terms, but it is often not necessary, *per se*.

It is worth mentioning that the modern QSAR paradigm extends the initial one proposed by Hansch in many directions: from congeneric to heterogeneous compounds, from single to multiple modes of actions, from linear regression to non-linear models, from simple to complex endpoints.

2.5 Human Toxicology and Drug Design

The discovery of new medical treatments is time consuming, and incredibly expensive. Drug discovery is the area in which chemoinformatics is routinely used. Drug discovery starts from the identification of a biological target that is screened against many thousands of molecules to identify the hits (molecules that are active). A number of those hits will produce a lead, a fragment that appears responsible for the wanted effects.

A lead has some desirable biological activity [22]: it is not extremely polar, does not contain toxic or reactive functional groups, has a small molecular weight and a low log P, has a series of congeners to allow structural modification. The leads are then combined with other elements to obtain the candidate drugs, in a process that requires multiple optimizations: reduced size, reduced toxicity, bioavailability.

Chemical space is the term given to the space that contains all of the theoretically possible molecules. However, when considering drug-like chemicals, the space becomes bounded according to known conditions such as the Lipinski rule-of-five [23] where a set of empirically derived rules is used to define molecules that are more likely to be orally available as drugs. This drug-like chemistry space is estimated to contain at least 10^{12} molecules [16], a very huge number.

To be able to explore this vast chemical space, it is necessary to deploy computer systems. There are two general approaches to drug design: one optimizes the molecule directly so to satisfy the binding for a particular target and the other one optimizes the molecule for a desired biological activity. The former is 3D based, while the latter uses only topological information. Often the two methods are integrated: using combinatorial chemistry and then QSAR, a small set of molecules with a high desired activity are selected. Then their shape is studied to see how they can fit the constraints of the binding site. In this last part, a new approach that simulates the binding using methods derived from robotics has also been developed [24].

Toxicity testing typically involves studying adverse health outcomes in animals administered with doses of drugs or toxicants, with subsequent extrapolation to expected human responses. The system is expensive, time-consuming, low-throughput, and often provides results of limited predictive value for human health. The toxicity testing methods are largely the same for industrial chemicals, pesticides and drugs, and have led to a backlog of more than 80,000 chemicals to which humans are exposed but whose potential toxicity remains largely unknown. This potential risk has urged national and international organizations in making a plan for assessing the toxicity of those chemicals.

In the USA, EPA (Environmental Protection Agency) routinely uses predictive QSAR based on existent animal testing to authorize new chemicals. Recently in the USA, a new toxicity testing plan, "Human Toxome Project", has been launched which will make extensive experimentation using predictive, high-throughput cell-based assays of human organs to evaluate perturbations in key pathways of toxicity. There is no consensus about this concept of "toxicity pathway" that in the opinion of many should be instead "disruption of biological pathways". The target of the project is to gain more information directly from human data, so to check in a future, with specific experiments, the most important pathways.

In the European Union, the REACH legislation for industrial chemicals has been introduced together with specific regulations for cosmetics, pesticide, food additives.

2.6 Environmental Toxicology

One of the most known episodes that draw the attention to the environmental pollution happened in the 1950s in Japan. Outbreaks of methylmercury poisoning occurred in several places in Japan due to industrial discharges of mercury into rivers and coastal waters. In Minamata bay alone, more than 600 people died. After that, mercury has been recognized as a pollutant and its presence in food monitored. In December 1970 a chemistry professor at New York University bought canned tuna and found a mercury dose 20 times higher than the limits of the FDA (Food and Drug Administration). It confirmed that the mercury poisoning was much more diffused, mainly in fish. Fishes have a natural tendency to concentrate mercury, especially the ones that are high on the food chain.

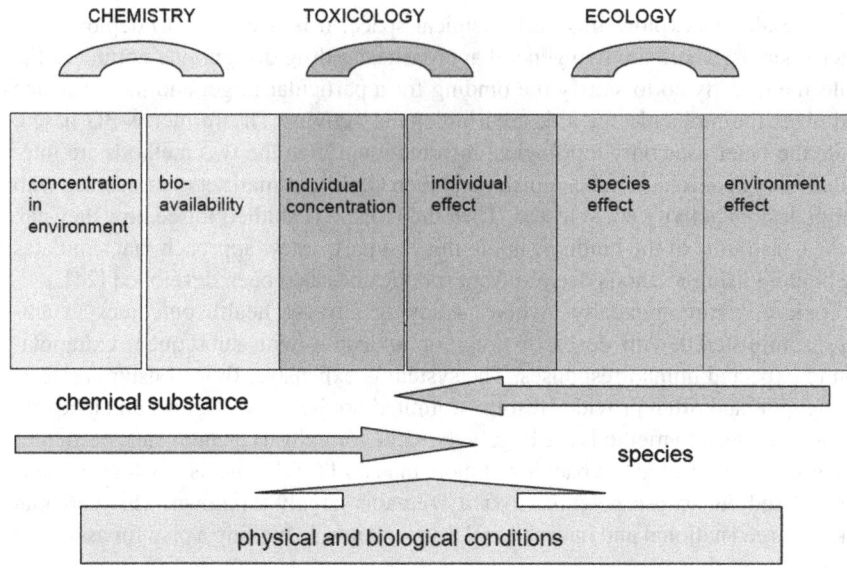

Fig. 2.6 Chemistry, toxicology, and ecology

The recognition that environment pollutants can harm humans in various ways, brought in the 1960s to the development of *environmental toxicology*. Environmental toxicology is concerned with the study of the harmful effects of various chemical, biological, and physical agents on living organisms. The steps of ecotoxicity studies are:

- entry, distribution, and fate of pollutants within the environment;
- entry and fate of pollutants in living organisms of the ecosystem;
- harmful effects on the constituents of ecosystems (which include humans).

Figure 2.6 indicates the main relationships between chemistry, toxicology, and ecology.

To address environmental toxicology some tests are used. Ideally a pollutant has to be tested on invertebrates and vertebrates, in air and in water species. Acute toxicity is usually the property studied on all those species, and *in silico* tools are applicable.

The regulations for human and environmental protection are out of the scope of this chapter. We only indicate that different regulations apply for air pollutants, industrial products (e.g., REACH), food, drinking water, cosmetics and detergents, pesticides, and drugs. There is only limited international agreement on the regulations and doses.

2.7 Problems

Of the many open problems in assessing toxicology using *in silico* models we discuss about a few points. The first one is the causal or mechanistic value of the QSAR equation. The QSAR for LD_{50}, for instance, does not have a simple interpretation in term of logic sentences. This will lead to the problem of mechanistic interpretation. Another point is about ethical issues. Is it really needed to make experiments on animals?

2.7.1 Logical and Probabilistic Knowledge

At the origin of any model there is a core hypothesis. In the case of QSAR for toxicology, we assume that the toxicity is related to the chemical structure:

$$Tox = f(Chem)$$

where Tox is toxicity, $f(\cdot)$ is a mathematical function and $Chem$ represents the chemical compound.

However, we have to better understand the implications and limits of this equation.

- From the classical work of toxicologists we know that the dose makes a compound toxic. Toxicologists have defined a kind of standardized effect, such as the dose which produces a given effect (e.g., death in 50% of the cells). For instance, chemical A will give the same toxic effect of chemical B using a dose double of that of chemical B; what changes is the dose, not the effect. Thus we can compare different chemicals only on the basis of their chemical nature, because we have defined a standard effect.
- We understand that different chemicals require different toxic doses to produce the same effect.
- The toxic effect refers to a cell or organism. Does this have an influence? If we consider LD_{50}, immediately we see that the same dose on 50% animals produces an effect which is opposite to that on the other 50%, because half die and half stay alive. The toxic effect is also dependent on the organism.
- From previous point we see that the basic QSAR equation, which appeared as a deterministic one, can be better considered from a stochastic point of view.
- It is well known that the chemical effect is mediated by processes, which can be, in many cases, unknown. Thus, Tox can be better described as

$$Tox = tox_1 + tox_2 + \cdots + tox_n$$

each of these factors tox_i describes a different process related to toxicity.

- The chemical part is actually much more complicated than a chemical formula; for instance, biochemical processes can transform the original compound in new compounds, more or less toxic than the original one.

The debate between models and their use can be taken at large, using the recent Internet posts of P. Norvig about N. Chomsky. Norvig [25] recalls a paper by L. Breiman [26] that presents two cultures: the data modelling culture, saying that nature can be described as a black box that has a relatively simple underlying model which maps input variables to output variables, and the algorithmic modelling culture, considering that nature cannot be described by a simple model, but proposing complex statistical and probabilistic models. People working in the algorithmic approach, as people using Artificial Intelligence (AI) techniques, aim at finding functions that map from input to output variables, but without expectation that the form of the function reflects the true underlying nature. None of the approaches can describe why something happens. The first since causality cannot be expressed in a purely statistical approach, the second since the underlining hypothesis of (the weak approach to) AI implies that the model simply emulates the effects of reality.

Breiman explains his objections to the data modelling culture. Data modelling makes conclusions about the model, not about reality. There is no way to uniquely model the true form of nature's function from pairs of inputs/outputs. What this model can do is to generalize to new data, not to give us the true form of a function. Whether this true form exist or not, it is not the task of a modelling and simulation method, but a matter of the right generalization process, something where humans are still superior to machines.

2.7.2 Mechanism or Causality

Hume argued that causality cannot be perceived and instead we can only perceive correlation. And indeed the basic biological experiments aim at finding a correlation (positive or negative) between some features and effects.

Biologists then want to understand why the effect can be explained in terms of metabolism, transformation substances, and so on. This is often called with the vague terms of "mode of action", or "mechanistic interpretation". Vagueness derives from the fact that there is no unique definition of mode of action: in some cases this is an observed behaviour as narcosis, in other cases it is a supposed chemical transformation. This is more complex than considering the organic chemical transformations since they happen in an organism where different biological pathways are usually supposed.

Inferring causality from data through Bayesian networks is today an active area of research and hopefully some answers could be found using those tools [27].

2.7.3 Ethical Issues

Toxicity testing typically involves studying adverse health outcomes in animals subjected to high doses of toxicants with subsequent extrapolation to expected human responses at lower doses. The system is expensive, time-consuming, low-throughput and often provides results of limited predictive value for human health. There are more than 80,000 chemicals to which humans are potentially exposed but whose potential toxicity remains largely unknown. Each year a few hundred new substances are registered. Is it really necessary to test all of them on animals?

The Declaration of Bologna, 1999, called the 3 R (Reduce, Refine, Replace), proposed a manifesto to develop alternative methods that could save millions of animals. In this scenario, the ethical issues are advocated also by authorities that have to protect humans, and that see use of animals as ethical than that of humans.

Generally, the stakeholders, often with competing needs, in the toxicity assessment are:

- Scientists and producers: they want modelling and discovery of properties. In other words, they want to build knowledge and translate it in products.
- Regulators and standardization organizations: they want be convinced by some general rule (mechanism of action). In other words, they want to reduce the risk of erroneous evaluations and be fast in decisions.
- Public, media, and opinion makers; they want to be fully protected against risk.

2.8 Conclusions

Since about 20 years chemical experimentation is more and more replaced by modelling and virtual experimentation. It has even been speculated that the vast majority of the discovery process for novel chemical entities will one day be performed *in silico* rather than in vitro or in vivo.

However *in silico* modelling of biological properties is a debated topic. Alongside classical methods as in vivo and in vitro experiments, the use of computational tools is gaining more and more interest. The usage of predictive QSARs is growing, since they provide fast, reliable, and quite accurate responses. They are candidates as accompaniment or replacement of existing techniques.

Finally, as shown in this chapter, the use of computers in chemistry and life sciences brings better tools to science and an open question: is computing (i.e., algorithms) able to capture and express knowledge about physical systems, and biological phenomena in particular?

References

1. Lynch M (2004) Introduction of computers in chemical structure information systems, or what is not recorded in the annals. In: Proceedings of 2002 conference on the history and heritage of scientific and technological, information systems, pp 137–148
2. Brown N (2009) Chemoinformatics—An introduction for computer scientists. ACM Comput Surv 41(2):8:1–8:38
3. Gasteiger J, Engel T (2003) Chemoinformatics: a textbook. Wiley-VCH, Weinheim
4. Willett P, Barnard J, Downs G (1998) Chemical similarity searching. J Chem Inf Comput Sci 38:983–996
5. Benfenati E, Gini G (1997) Computational predictive programs (expert systems) in toxicology. Toxicology 119:213–225
6. Gini G, Katritzky A (Eds) (1999) Predictive toxicology of chemicals: experiences and impact of Artificial Intelligence tools. In: Proceedings of AAAI spring symposium on predictive, toxicology, SS-99-01
7. Hartung T (2009) Toxicology for the twenty-first century. Nature 460(9):208–212
8. Scerri E (2006) The periodic table: its story and its significance. Oxford University Press, New York
9. Balaban A (1985) Applications of graph theory in chemistry. J Chem Inf Comput Sci 25:334–343
10. Weininger D (1988) SMILES, a chemical language and information system. Introduction to methodology and encoding rules. J Chem Inf Comput Sci 28:31–36
11. Morgan HL (1965) The generation of a unique machine description for chemical structures—A technique developed at chemical abstracts service. J Chem Docum 5:107–113
12. Gillet V, Willet P, Bradshaw J, Green D (1999) Selecting combinatorial libraries to optimize diversity and physical properties. J Chem Inf Comput Sci 39:169–177
13. Chow P, Ng R, Ogden B (eds) (2008) Using animal model in biomedical research. World Scientific Publishing Co, Singapore
14. Livingston D (2000) The characterization of chemical structures using molecular properties. A survey. J Chem Inf Comput Sci 40:195–209
15. Hansch C, Malony P, Fujita T, Muir R (1962) Correlation of biological activity of phenoxyacetic acids with hammett substituent constants with partition coefficents. Nature 194:178–180
16. Ghose A, Crippen G (1986) Atomic physicochemical parameters for three-dimensional structure directed quantitative structure-activity relationships I. Partition coefficients as a measure of hydrophobicity. J Comp Chem 7:565–577
17. Karelson M (2000) Molecular descriptors in QSAR/QSPR. Wiley-VCH, Weinheim
18. Hastie T, Tibshirani R, Friedman J (2001) The elements of statistical learning: data mining, inference, and prediction. Springer, New York
19. Ferrari T, Gini G, Golbamaki Bakhtyari N, Benfenati E (2011) Mining structural alerts from SMILES: a new way to derive structure-activity relationships. In: Proceedings of 2011 IEEE CIDM, pp 120–127
20. Gini G, Benfenati E (2007) E-modelling: foundations and cases for applying AI to life sciences. Int J Artif Intell T 16(2):243–268
21. Ashby J (1985) Fundamental SAs to potential carcinogenicity or noncarcinogenicity. Environ Mutagen 7:919–921
22. Jorgensen W (2004) The many roles of computation in drug discovery. Science 303:1813–1818
23. Lipinski C, Lombardo F, Dominy B, Feeney P (2001) Experimental and computational approaches to estimate solubility and permeability in drug discovery and development settings. Adv Drug Deliv Rev 46:3–26
24. Cortes J, Jaillet L, Simeon T (2007) Molecular disassembly with RRT-like algorithms. In: Proceedings of 2007 IEEE ICRA, pp 3301–3306

25. Norvig P (2012) http://norvig.com/chomsky.html, accessed July 2013
26. Breiman L (2001) Statistical modelling: the two cultures. Stat Sci 16(3):199–231
27. Kalisch M, Mächler M, Colombo D, Maathuis M, Bühlmann P (2012) Causal inference using graphical models with the R package pcalg. J Stat Softw 47(11):1–26

Part II
From Science to Computing

The longstanding debate on the scientific nature of computing is a leitmotif along the all history of the discipline. Notwithstanding the different positions expressed, the "scientificity" of computing, even of a peculiar type, has been almost universally recognized both inside and outside the community. Part II of this volume discusses on the nature and role of a central scientific concept, such as the experimental method, within computing, and in particular within a particularly interesting area of computer engineering that is autonomous robotics, which features large uncertainties due to the interactions of robots with the real world. While recognizing the importance of adopting an experimental approach, especially in evaluating the artifacts produced, rigorous experimental practices seem not to be fully adopted and applied through the field. The attention to experimental practices of other scientific disciplines could represent an important reference point in the assessment of disciplinary nature of autonomous robotics and, more generally, of computing. Along this path, precisely assessing how experiments are performed in the current practice seems a sound initial step. This is the topic of the following chapter. However, as discussed in the second chapter of this part, the practical application of the experimental method to autonomous robotics raises some difficulties, which show the necessity for computing to reflect on its methodology, given some peculiarities of the field with respect to other more traditional sciences.

Chapter 3
Good Experimental Methodologies for Autonomous Robotics: From Theory to Practice

Francesco Amigoni, Viola Schiaffonati and Mario Verdicchio

Abstract A lively discussion on good experimental methodologies has recently taken place in the autonomous robotics community. Workshops have been organized, special issues published, and projects funded after recognizing that experimentation in autonomous robotics has not yet reached a level of maturity comparable with other fields of engineering and science. Within this discussion, we are not aware of any systematic survey on how experiments are conducted in autonomous robotics papers published in major journals and conferences. In this chapter, we provide an initial contribution to fill this gap by analyzing the experimental trends that emerge from the autonomous robotics papers presented over the last 11 years at the International Conference on Autonomous Agents and Multiagent Systems (AAMAS), which constitute a privileged sample to get a picture on the evolution of experimental activity in the area. We conduct our analysis in the light of some principles that have been proposed for the development of good experimental methodologies in autonomous robotics.

Keywords Autonomous robots · Experimental methodologies

F. Amigoni (✉) · V. Schiaffonati
Dipartimento di Elettronica, Informazione e Bioingegneria, Politecnico di Milano, Milan, Italy
e-mail: francesco.amigoni@polimi.it

V. Schiaffonati
e-mail: viola.schiaffonati@polimi.it

M. Verdicchio
Dipartimento di Ingegneria,
Università degli Studi di Bergamo, Bergamo, Italy
e-mail: mario.verdicchio@unibg.it

F. Amigoni and V. Schiaffonati (eds.), *Methods and Experimental Techniques in Computer Engineering*, PoliMI SpringerBriefs,
DOI: 10.1007/978-3-319-00272-9_3, © The Author(s) 2014

3.1 The Gap Between Theory and Practice

A lively discussion on good experimental methodologies has recently taken place in the autonomous robotics community. Workshops have been organized at major conferences [1], special issues published [2], and projects funded [3–6], which reflect the recognition that experimentation in autonomous robotics has not yet reached a level of maturity comparable with other fields of engineering and science. The interest in experimental methodologies is also motivated by the idea that they could reduce the gap between industrial applications and those applications, like service robotics, that require a significantly higher level of autonomy. If industrial robotics has some standards to evaluate systems employed in factories, the role of standardized experimental evaluation in autonomous robotics has not yet been universally recognized [7]. Within the context of this discussion, we are not aware of any systematic survey of the experimental activities presented in autonomous robotics papers published in major journals and conferences.

In this chapter, we aim at contributing to fill this gap by considering the autonomous robotics papers presented at the International Conference on Autonomous Agents and Multiagent Systems (AAMAS) over the last 11 years (from the inaugural edition in 2002 to the one in 2012). The selection of this source is motivated by a number of factors. Firstly, AAMAS is obviously a showcase of autonomous robotics research and features papers by many authors who regularly publish also at International Conference on Robotics and Automation (ICRA), International Conference on Intelligent Robots and Systems (IROS), and other robotic venues. However, differently from ICRA and IROS, autonomous robotics papers are well identified: they are usually presented in dedicated sessions. Moreover, considering papers from AAMAS is interesting to account for the point of view on experiments of researchers at the intersection between robotics and autonomous agents, which is a privileged perspective for observing autonomous robotics.

In this chapter, we analyze the trends that emerge from the experimental activities reported in 95 robotics papers presented at AAMAS in the light of some principles that have been proposed for the development of good experimental methodologies in autonomous robotics [8]. Although we are not claiming that they are the only principles that should be adopted in defining experimental methodologies for autonomous robotics, we deem that they are at the very foundation of experimental activity as traditionally intended and, hence, cannot be ignored. These principles are summarized below.

Comparison. The meaning of comparison in science is twofold: in a wider scope in the literature, it means to know what has already been done within the field, to avoid the repetition of uninteresting experiments and to get hints on promising issues to tackle; in a narrower scope of a specific kind of experiment, it refers to the possibility for researchers to accurately compare new results with old ones.

Reproducibility and repeatability. These features are related to the very general idea that scientific results should undergo to the most severe criticisms in order to be confirmed: reproducibility is the possibility for independent scientists to verify the results of a given experiment by repeating it with the same initial conditions, instruments, and techniques, whereas repeatability is the property of an experiment that yields the same outcome from a number of trials, performed at different times and in different places.

Justification and explanation. This principle deals with well-justified conclusions based on the information collected during an experiment: it is not sufficient to collect as many precise data as possible, but it is also necessary to look for an explanation, that is, all the experimental data should be interpreted in order to derive the correct implications that lead to the conclusions.

Starting from the above principles, we identify a number of issues on how experimental activities are conducted. These issues include whether the papers present experiments, whether the experiments are performed in simulation or with real robots, whether the data or the code is available, whether the proposed system is compared with alternative systems, and so on. To the best of our knowledge, no other work has provided such an extensive survey over experimental activities in autonomous robotics. Our aim is not to present new methodologies, but to critically discuss some emerging trends that could help shed some light on the future of experiments in this field.

This chapter is organized as follows. The next section introduces the methodology we have used to select and analyze the papers we surveyed. Section 3.3 reports on the results of the analysis, while Sect. 3.4 discusses these results in the light of the experimental principles listed above. Finally, Sect. 3.5 concludes the paper.

3.2 Methodology

In this section we present the criteria we adopted in the process of selecting and analyzing the papers.

From the span of AAMAS conferences, from 2002 to 2012, we have considered all the papers that have been classified into robotics-related sessions. In this way, we avoided any arbitrary decisions in the selection of the papers. The titles of the sessions and the corresponding number of papers are reported in Table 3.1. In 2004 and 2007 no specific sessions on robotics were included in the proceedings. We ignored a 2007 session titled "Embodied agents and architectures", because it contains mostly papers related to synthetic and emotional agents. Overall, 95 papers have been analyzed. We explicitly note that these papers can be univocally identified in the AAMAS proceedings[1] starting from the information contained in Table 3.1.

A preliminary analysis of posters and short papers showed that there is no significant difference in the quality of the reported experimental activities with respect to full papers, although the number of experiments presented is obviously limited due to space constraints. Thus, since we are interested in analyzing the quality and not the quantity of experiments, posters and short papers have been included in our survey. Table 3.2 shows a classification of the 95 papers with respect to some of the keywords listed in the AAMAS2013 call for papers.

---

[1] Available at http://www.ifaamas.org/.

Table 3.1 Robotics papers at AAMAS

Year	Session	Number of papers
2002	Mobile embodied agents	9
	Robot architectures	4
2003	Robotics	4
2005	Robotics	5
	Robotics posters	3
2006	Robotics	10
2008	Multi-robotics full	8
	Multi-robotics short	11
2009	Multi-robotics	6
2010	Robotics I	6
	Robotics II	6
2011	Robotics	5
	Robotics and learning	4
2012	Robotics I	5
	Robotics II	5
	Robotics III	4

Table 3.2 Topics of robotics papers at AAMAS

Topic	Number of papers
Integrated perception, cognition, and action	12
Machine learning for robotics	10
Mapping, localization and exploration	6
Robot planning	32
Robot teams, multi-robot systems	5
Robot coordination	21
Robotic agent languages and middleware for robot systems	8
E-learning	1

The 95 papers we considered usually present techniques and methods that are applied to specific applications and problem settings, as shown in Table 3.3. Beyond generic navigation tasks that are fundamental for autonomous mobile robots, recurring applications include robotic soccer, surveillance and security (e.g., target recognition and tracking in military settings), and group behaviors like flocking. Other applications, aggregated in the final row of Table 3.3, include damage detection, self-assembling, robot dancing, and visitor companion (and e-learning). Note that two papers of 2005 and 2008, respectively, address two applications each.

We have considered a number of issues on how experimental activities are conducted to assess the properties of the proposed autonomous robotic systems. These issues are general enough to be applied to the wide range of topics the surveyed papers deal with. The list below is not definitive nor exhaustive, but it reflects our understanding of the issues characterizing experimental activities at the intersection between robotics and autonomous agents.

Table 3.3 Applications of robotics papers at AAMAS

Topic	Number of papers	2002	2003	2005	2006	2008	2009	2010	2011	2012
Environmental monitoring	2					1			1	
Exploration	7	1		1		1	1		1	2
Flocking and formation	11					5	1	2		3
Foraging	3			1		1				1
Generic navigation (e.g., coverage)	24	2		1	3	4	2	5	4	3
Locomotion (e.g., walking)	3	1	1		1					
Manipulation	2					1			1	
Object transportation	4	2	1	1						
Rescue	3					1		1	1	
Robotic soccer	12	2			3	4			1	2
Surveillance (e.g., patrolling and pursuit)	10	1		2		2	1	3		1
Target recognition and tracking	9	2	1	2	3					1
Other applications	7	2	1	1			1	1		1

Experiments. Does the paper present any experimental activity? By *experiments* we mean activities that require the implementation and the run of a computing system. In this sense, simple illustrative examples that are "drawn on paper" are not considered experiments for the purposes of this work. Moreover, we considered only what is reported in the paper: a paper does not qualify as a work that presents experiments if the only experimental activities discussed are illustrated in another work or are only claimed without explicit results.

Simulation or real robots. Among the papers describing experiments according to the above-mentioned criterion, we distinguish between simulations and activities with real robots. We further make a distinction between *standard* and *custom platforms*, that is, commercial and publicly available platforms, and the ones that are usually not made available outside the lab that developed them, respectively.

Data/code availability. Does the paper make *data* and *code* regarding the presented experiments available (e.g., downloadable from a website)?

Comparison. By *comparison* with other systems, we mean whether the proposed system is experimentally compared with others that perform the same function. We further distinguish between systems compared with *variants* developed by the same authors or with baseline systems, and systems compared with *alternatives* developed by other authors. An example of the former case is a learning system evaluated with or without a deadline for learning, while the latter case includes comparing the performance of the proposed learning system with that of Q-learning, for example. Please note that we consider comparisons with baseline systems, like random methods in case of task assignment, as comparisons with variants developed by the same authors.

Measures. We check what is *measured* in the experiments, with a classification into two broad categories: *effectiveness* (or *functional*) measures, i.e., measures of what a system is supposed to do, and *efficiency* (or *non-functional*) measures, i.e., measures of the resources the system requires to obtain such performance. Examples of functional measures include the rewards obtained by robots in a reinforcement learning setting, the probability of detecting an intruder in a patrolling application, and success rate in reaching some desired configuration for swarm robotic systems. Examples of measures that we consider non-functional are the communication overhead associated to a given performance, the rate of convergence for a learning algorithm, and all measures of time and space complexity.

Settings. We consider the *settings* in which experiments are performed, and make a distinction between experiments performed in a single setting (e.g., one environment for a navigation algorithm or one task for a learning algorithm) and in different settings (e.g., multiple environments or multiple tasks).

Statistical analysis. Finally, we focus on the way in which the experimental activities reported in papers deal with *uncertainty* and *randomness* that affect all robotic systems operating in dynamic environments: are papers just presenting averages (or medians) or are they also presenting a more robust statistical analysis (e.g., based on ANOVA)? We consider papers showing at least standard deviation as presenting (a very simple, indeed) statistical analysis.

We will discuss in Sect. 3.4 how these issues are related to the experimental principles mentioned in Sect. 3.1.

As in any survey, our selection of issues is subjective and amendable. However, the issues we consider are so general that they can be applied to all the topics in Table 3.2. Moreover, our analysis of the papers presents some inescapable limitations. For example, sometimes multiple systems are presented in the same paper, that could be classified in different ways: some of them are compared with alternatives, whereas some others are not. In such cases, we chose to go for the best available option (e.g., we considered the whole paper as comparing the proposed system with alternatives if it is done for at least one of the proposed systems). Notwithstanding its limits, we believe that the analysis reported in the next section contributes to form an initial picture of the experimental trends in autonomous robotics.

3.3 Analysis

In this section, we present the results of our analysis of the AAMAS robotics papers and try to identify some emerging trends in the experimental activities of autonomous robotics. The results of our analysis are presented in graphs, showing the fraction of papers addressing the issues illustrated in the previous section.

Let us start from Fig. 3.1, which shows that the majority of papers present experiments. This should not come as a surprise, as experimentation is the main way in

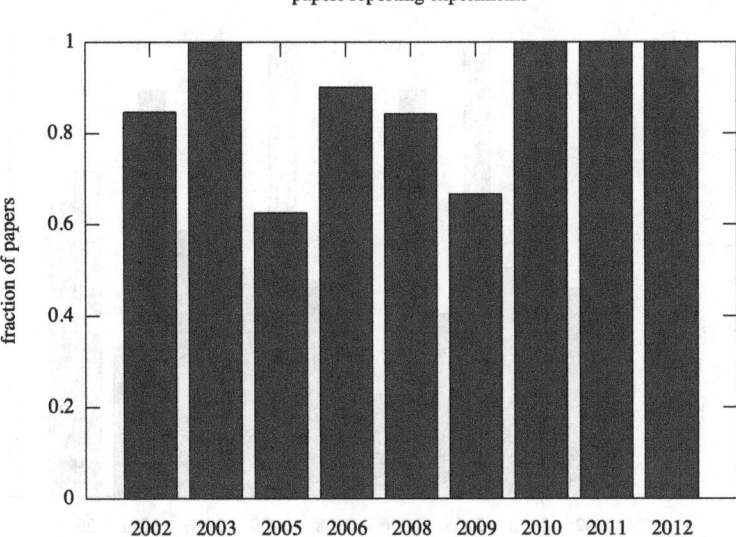

Fig. 3.1 Fraction of AAMAS robotics papers that present experiments

which robotic systems are evaluated and assessed. In some years of the surveyed time span, all papers present experimental activities, whereas in some others such fraction is lower. For example, in 2005, 3 out of 8 papers do not present experiments; 2 (short, 2-page) papers claim that experiments have been performed, but do not describe them, while another (full) paper is theoretical and presents only simple examples of application of the proposed algorithm for pursuit evasion. It is interesting to notice that all papers in the last 3 years (2010, 2011, and 2012) present experiments.

Obviously, the way in which experiments are intended strongly varies (also according to the topic of the paper): they range from simple qualitative descriptions of the behavior of the implemented systems to more sophisticated evaluation activities, involving different alternative approaches and possibly supported by statistical analysis. Let us then try to shed some light on how experiments are performed.

Figure 3.2 shows the fraction of papers with experiments that use simulated and real robots. Note that a paper may present experiments with both simulated and real robots. In such a case, the paper contributes to both fractions, which is why the bars relative to a year may sum up to more than 1. Simulation dominates over real robots, which can be explained by the lower costs and the relatively easier operational aspects of simulation. However, it is interesting to notice that the fraction of papers presenting experiments with real robots is somehow constant over the years. This could be related to the fact that papers addressing some common topics, like target tracking, are more frequently presenting experiments with real robots. A common

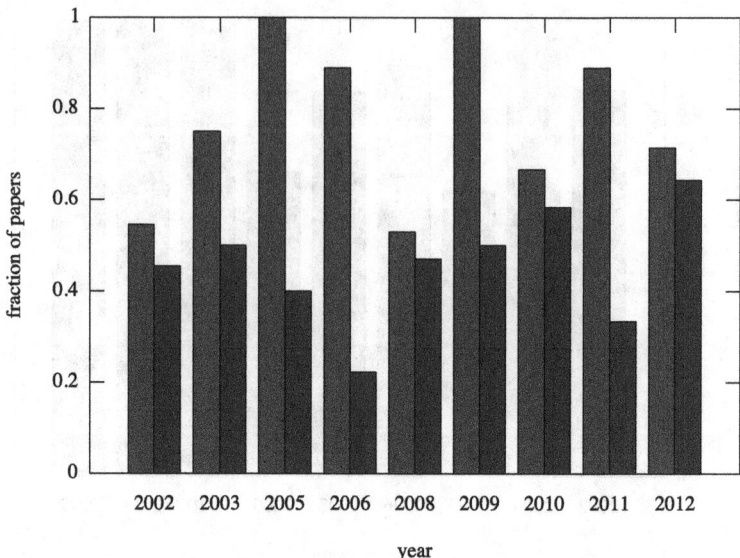

Fig. 3.2 Fraction of papers that use simulated and real robots in experiments

Table 3.4 Standard simulators used in robotics papers at AAMAS

Simulator	Number of papers	2002	2003	2005	2006	2008	2009	2010	2011	2012
(Player/) Stage or Gazebo	4				1				1	2
Robocup simulators (e.g., USARSim)	6				1	1		1	2	1
Cyberbotics' webots	6	2			1	1		1	1	

situation is that in which extensive experiments are performed in simulation and simpler demonstrations are performed with real robots.

Figure 3.3 shows the analysis of the simulation tools, with a particular focus on standard simulators as opposed to custom ones. The use of standard simulators seems to be increasing over the years, which could be related to the fact that more and more reliable simulation platforms have recently become available. Looking at the standard simulators used in the last years, it emerges that most of them are used in competitions like RoboCup (e.g., USARSim [9]). Another standard simulator that has been used since 2002 and is still employed is Cyberbotics' Webots [10]. The standard simulators that have been used at least by 2 papers are reported in Table 3.4.

The trend of using standard platforms is even stronger in the case of experiments with real robots: Fig. 3.4 shows that there is an evident tendency in adopting standard

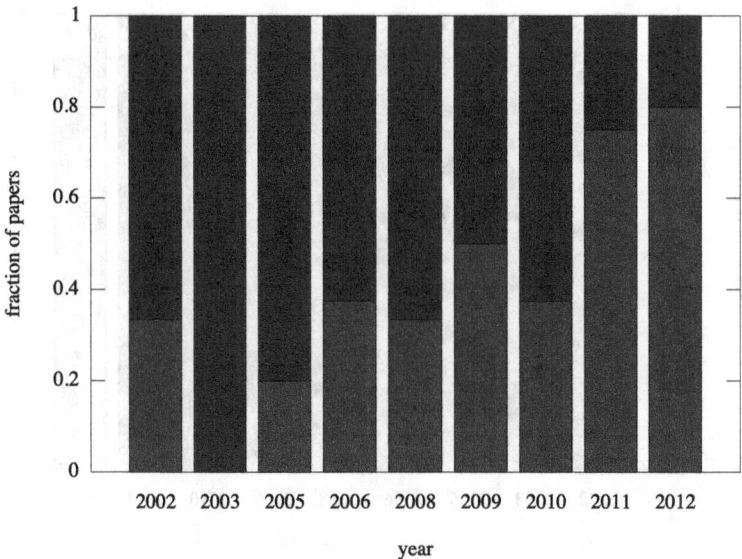

Fig. 3.3 Fraction of papers that employ standard and custom simulators

platforms in this type of research. A possible explanation is that standard robotic systems are usually easier to set up, more reliable, and, in some cases, cheaper in terms of resources and time than custom systems. Among the standard platforms that are employed more often, there are MobileRobots' Pioneer robots [11] and Sony's Aibo robots [12]. The standard robotic platforms that have been used at least by two papers are reported in Table 3.5.

Figure 3.5 shows the fraction of the papers that, whether with simulations or with real robots, compare their proposal with other systems. This analysis is interesting because it shows that, in a number of cases, robotic systems presented at AAMAS are only empirically shown to work, but such experiments do not include a comparison with other systems, which in some sense weakens their assessment.

Among those papers that provide experimental comparisons, some take only simple variants of the proposed systems into account, whereas other papers consider fully alternative systems, typically developed by other researchers. Figure 3.6 shows these two trends: it is evident that there is an increasing tendency toward a more sophisticated notion of comparison.

Figure 3.7 shows the fraction of papers (among those that present experiments) presenting functional (effectiveness-related) and non-functional (efficiency-related) experimental measures. If functional measures are more frequently presented (not surprisingly, since the goal of experimenting is to show that a proposed system works), the illustration of non-functional measures has reached a good diffusion in the last years. This allows for a more complete understanding and assessment of

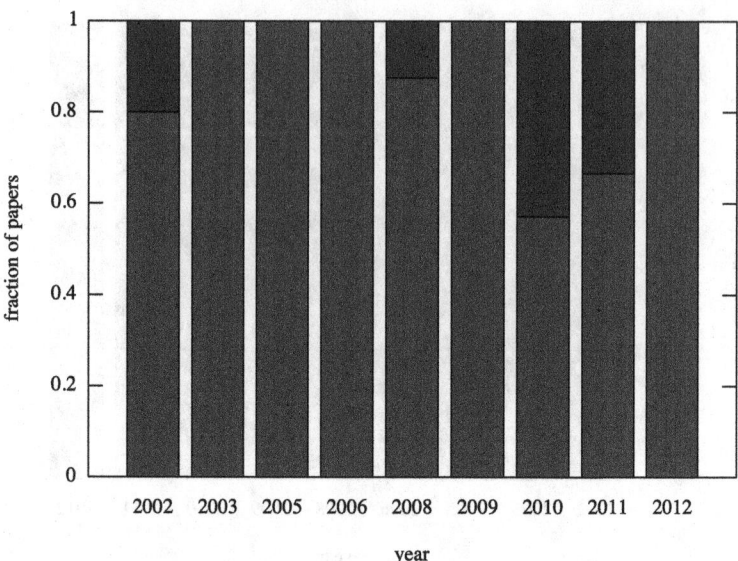

standard (blue) or custom (red) real robotic systems

Fig. 3.4 Fraction of papers that employ standard and custom real robotic systems

Table 3.5 Standard real robotic systems used in robotics papers at AAMAS

Robotic system	Number of papers	2002	2003	2005	2006	2008	2009	2010	2011	2012
Sony's AIBO	2					2				
KOVAN's kobot	2						1	1		
Lego's mindstorms	2	1		1						
Mobilerobots' pioneer family (I, 2-DX, 3-DX, P3-AT)	8	2	2		1		1	1		1
Friendly robotics' RV400	3					1			2	
Segway's RMP	2			1		1				
Albedaran's nao	3									3
EPFL's e-puck	2									2

robotic systems, by showing not only the success rate in the tasks a system has been designed for, but also the resources that are needed to achieve such results.

Figure 3.8 shows the papers (among those presenting experiments) that illustrate experiments in single or multiple settings. Naturally, an experimental activity conducted in multiple settings is expected to come to stronger conclusions than an experimental activity conducted in a single setting. However, sometimes the use of a single

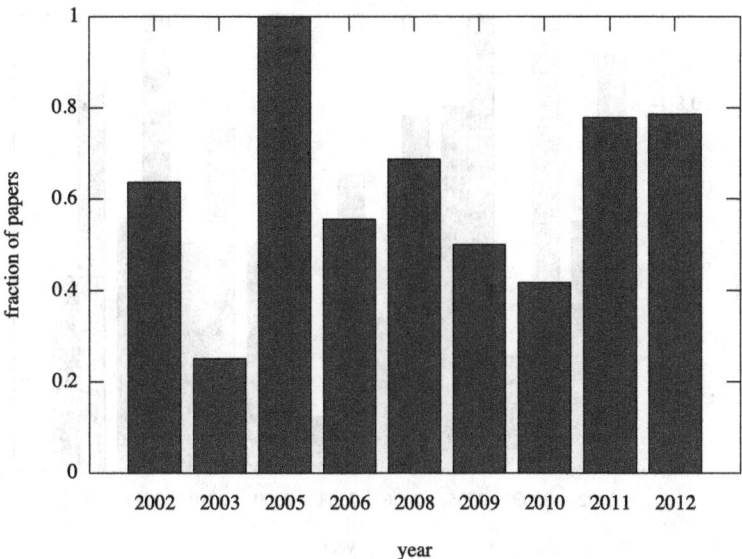

Fig. 3.5 Fraction of papers that compare the proposed systems with other systems

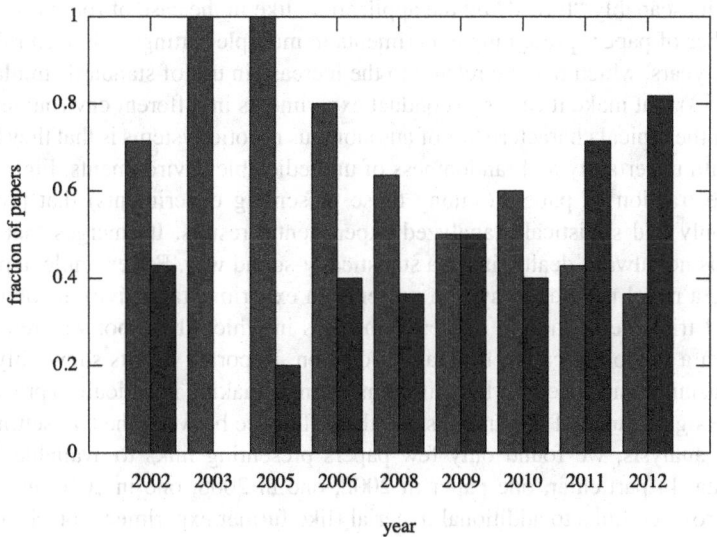

Fig. 3.6 Fraction of papers that compare the proposed systems with variants of the same systems or with alternative systems

functional (blue) or non-functional (red) measures

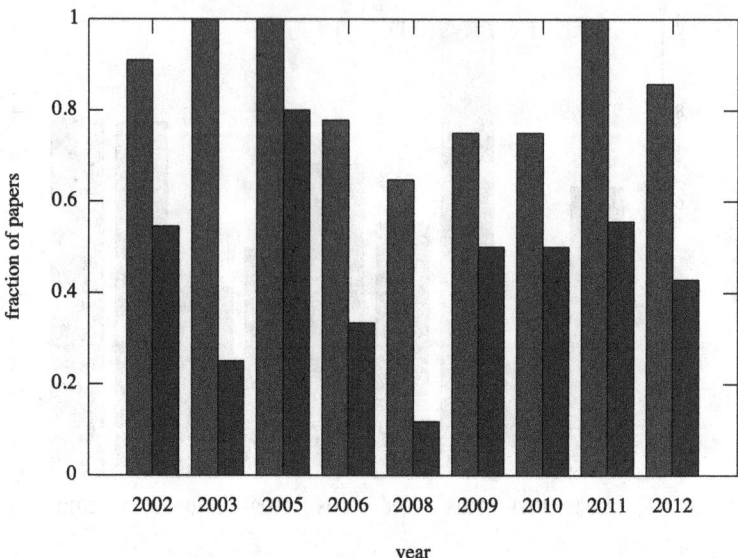

Fig. 3.7 Fraction of papers that present functional and non-functional experimental measures

setting is inescapably "forced" by the application, like in the case of robotic soccer. The number of papers presenting experiments in multiple settings is indeed raising in the last years, which may be related to the increase in use of standard simulators (see Fig. 3.3) that make it easier to conduct experiments in different environments.

One of the typical characteristics of autonomous robotic systems is that they have to deal with uncertainty and randomness of unpredictable environments. Figure 3.9 shows the fraction of papers (among those presenting experiments) that present average-only and statistically analyzed experimental results. It emerges that randomness is not always dealt with in a statistically-sound way. For example, a paper proposing a novel navigation system presents an experimental activity involving a number of trials performed in real environments in which the robot was required to go from a random location to a target location. Reported results show only the average distance and speed for two different settings, making it difficult to precisely assess the significance of the findings and the difference between the two settings.

In our analysis, we found only few papers presenting links to available code and/or data. In particular, one paper in 2002, one in 2008, one in 2010, and two in 2012 provided links to additional material (like further experiments or videos of experiments), but the links appear to be broken. Other papers, however, i.e., one in 2003, two papers in 2008, three in 2010, and five in 2012, featured links to additional material which are working (as of April 2012).

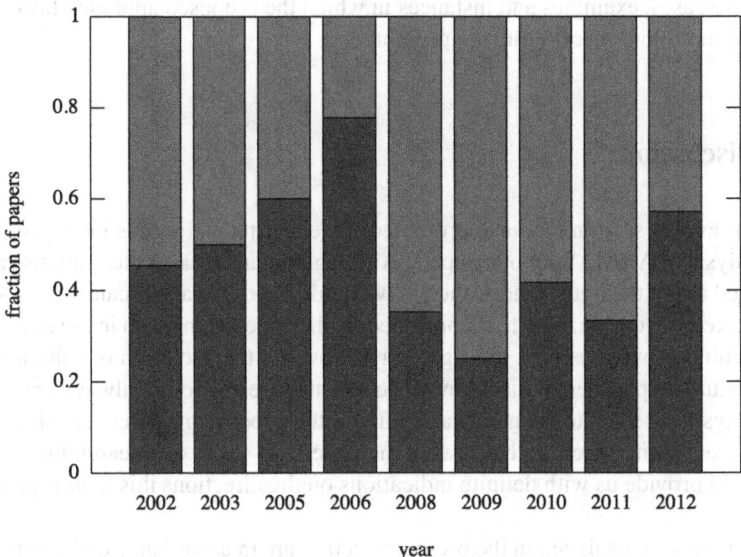

Fig. 3.8 Fraction of papers that present experiments in single or multiple settings

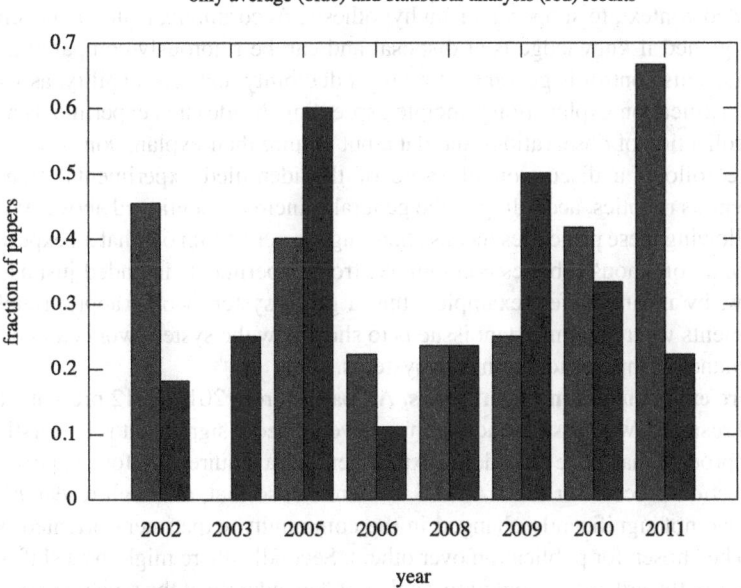

Fig. 3.9 Fraction of papers that present average-only and statistically-analyzed experimental results
(note truncated scale on vertical axis)

Finally, we observed that very few papers (5 out of 81) report negative results. In these rare cases, examples and instances in which the proposed approach fails or for which it has a low success rate are presented.

3.4 Discussion

In the previous section we have presented the experimental trends emerging from our analysis of AAMAS robotic papers. Although not in all cases clear trends can be identified out of the figures, nevertheless we think that our analysis can point at some clues to keep an eye on, which, if confirmed by the papers published in the upcoming years, could provide us with the right perspective on the evolution of robotics as a more mature experimental discipline. The fact that trends do not always emerge in clear ways can be due to the great variability of the papers considered and also to the limited pool from which we have taken the papers: 11 years of research may not be enough to provide us with definite indications on the directions this field is going to take.

The issues we analyzed in the previous section are in accordance to the very idea of experiment as developed in science after the so called Scientific Revolution in the seventeenth century, and based on the principles listed in Sect. 3.1. An experiment is a *controlled experience*, namely a set of observations and actions, performed in a controlled context, to support a given hypothesis. Accordingly, better achievements can be gained if knowledge is at disposal and can be rigorously compared among scholars. This control is guaranteed by reproducibility and repeatability, as well as by the justification/explanation principle expressing the idea that experiments are not just a collection of observations and data, but require their explanation.

Here follows a discussion of some of the identified experimental trends in autonomous robotics, according to the general principles mentioned above. We note that following these principles means changing the perception of what an experiment is in the autonomous robotics community: from experiments intended just as ways to show, by means of few examples, that a given system works appropriately, to experiments where an important issue is to show how the system works and how its performance compares to alternative systems.

More experiments in recent years. All papers from 2010–2012 present experimental results. Two possible factors may have played a significant role. Firstly, the review process may have considered experiments as a requirement for a paper to pass the selection process for the AAMAS conference, so that, although research practices have not significantly changed in the community, experiment-oriented works tend to be chosen for publication over others. Secondly, there might be a shift in the community toward experimentation, so that, independently of the reviewing process, all in all, we have more experimentation in the field of robotics.

More simulation than real robots. Beyond the simpler explanation that simulations are cheaper than physical robots, this tendency might depend on the fact that simulators have become more and more sophisticated, enabling researchers to

model robots and environments with more detail and precision. The increased reliability of the simulation results may have triggered a change in the attitude of robotics
researchers toward simulators, not seen as "fake robotics" anymore, but as first-class
citizens of the field.

Increased use of standard platforms. Adopting a well-documented, existing
tool is often a quicker and easier way to obtain results than starting from scratch.
The above-mentioned increased reliability of simulators, and a similar phenomenon
in the context of physical robots may have boosted the adoption of standard systems
throughout the research community. This trend is also supported by efforts like
ROS [13], which aims at providing a common platform for the control modules of
the robots. The use of standard simulators and robots has surely a positive effect on the
reproducibility and repeatability of experiments, so that they can be performed under
different conditions and by several researchers, with the possibility to strengthen
previously obtained results, or to put them under discussion.

Weakness of experimental comparison of systems. The fraction of papers that
feature an experimental comparison of the proposed system with other systems is not
very large. This phenomenon may be due to the difficulty to determine a common
ground for experimental comparisons. For instance, identifying a set of standard
environments for benchmarking navigation systems is an hard problem, let aside the
identification of measures for comparing the systems. However, although the number
of papers performing comparisons is not large, we have witnessed an increase in the
quality of comparison, including more frequent evaluation of alternative systems
developed by other researchers. This could be again connected to the adoption of
standard platforms. For example, using RoboCup soccer simulators, it is easy to
compare different teams by having them play against each other. Should this tendency
be confirmed in the future, it would surely have a positive impact on the scientific
aspects of autonomous robotics research, enhancing the comparison principle.

More attention toward non-functional parameters. Industrial robotics has had
a tremendous impact on the industry and the economy, whereas more cutting-edge
proposals coming from artificial intelligence and autonomous robotics have only
recently found their way into the world outside the labs (e.g., with entertainment
robots, household robots, prosthetics). This means that more and more products
of research in robotics must meet tighter requirements given by the services these
artifacts are supposed to provide in everyday life, which may have triggered in
researchers a bigger interest in the efficiency of their systems. Moreover, a better understanding of the systems strengthens the justification/explanation principle
toward a more scientific approach to experiments.

Little attention to statistical analysis. This trend evidences still a distance from
the principle of justification/explanation mentioned above. A rigorous statistical
analysis nowadays is a necessary requirement to treat data in a meaningful way
if we want to look for real explanations instead of mere collection of data. A similar
situation has been faced some years ago in artificial intelligence [14] and brought to
a widespread use of statistical analysis in many areas of the field.

Low availability of data and code. This trend, again, is not in accordance with
the general idea of experiments, according to which data should be compared in the

most rigorous and extensive way. In the papers we have analyzed, very few present full (raw) experimental data or make code available to the reader. Although in some cases there are copyright and privacy issues that prevent distribution of data and code, this shortcoming obviously weakens the possibility to replicate experiments by other researchers and hence to extensively control experimental results in an independent way. A possible way to overcome this problem could be the creation of a repository for additional material of the papers. Moreover, it could be useful to stimulate the use of publicly available datasets, like Radish [15] and RAWSEEDS [4], in experiments. A step in this direction could also address the complaint (sometimes encountered in AAMAS papers) that there are no standard benchmarks for evaluating autonomous robotic systems.

3.5 Conclusions

Our efforts in analyzing the AAMAS robotics papers of the last 11 years allowed us to spot some clues that make us optimistic about the future development of the experimental methodologies in this discipline. The increase in the use of standardized platforms, both for simulations and for real robots, seems to be the most promising of all. The convergence of the endeavors of several researchers toward the same platforms can have a positive impact on the significance of the results of their work, in that, they can be throughly verified under different conditions by different testers thanks to the standard systems that provide an easily accessible common ground to all the members of the community. Moreover, as simulators become more and more reliable, the financial constraints that come with physical robotic systems can be bypassed without compromising the verisimilitude of the experiments, thus allowing also researchers from smaller labs to join forces in the robotics enterprise. However, some other aspects seem more problematic and suggest that things should be done differently, like the scarce use of statistical analysis and the difficulty in making data and code publicly available to ease the comparison of systems.

We discussed only some trends that emerge from our analysis. Researchers can easily identify several other patterns that it is worth discussing. Future work could address a more complete analysis of how experiments are conducted in autonomous robotics, both considering more issues (possibly specialized for particular sub-fields) and a larger sample of papers. In particular, it could be interesting to compare the trends emerged from our analysis of AAMAS robotics papers to those of other conferences, like ICRA. To this end, a criterion to select autonomous robotics papers from the large number of papers in the ICRA proceedings is needed.

References

1. Euron GEM Sig (2007) http://www.heronrobots.com/EuronGEMSig/. Accessed July 2013
2. Madhavan R, Scrapper C, Kleiner A (2009) Special issue on characterizing mobile robot localization and mapping. Auton Robot 27(4):309–481
3. BRICS (2009) Best practice in robotics. http://www.best-of-robotics.org/. Accessed July 2013
4. Rawseeds (2006) http://www.rawseeds.org/. Accessed July 2013
5. RoCKIn (2013) Robot competitions kick innovation in cognitive systems and robotics. http://rockinrobotchallenge.eu/. Accessed July 2013
6. RoSta (2007) Robot standards and reference architectures. http://www.robot-standards.eu/. Accessed July 2013
7. Bonsignorio F, Hallam J, del Pobil AP (2007) Good experimental methodology - GEM guidelines. http://www.heronrobots.com/EuronGEMSig/downloads/GemSigGuidelinesBeta.pdf. Accessed July 2013
8. Amigoni F, Reggiani M, Schiaffonati V (2009) An insightful comparison between experiments in mobile robotics and in science. Auton Robot 27(4):313–325
9. USARSim (2011) Unified system for automation and robot simulation. http://usarsim.sourceforge.net. Accessed July 2013
10. Webots (2011) http://www.cyberbotics.com/overview. Accessed July 2013
11. Adept Technology (2011) http://www.mobilerobots.com/. Accessed July 2013
12. Sony Aibo (2006) http://support.sony-europe.com/aibo/. Accessed July 2013
13. ROS (2011) Robot operating system. http://www.ros.org/. Accessed July 2013
14. Cohen P (1995) Empirical methods for artificial intelligence. The MIT Press, Cambridge
15. Howard A, Roy N (2003) The robotics data set repository (radish). http://radish.sourceforge.net/. Accessed July 2013

Chapter 4
RAWSEEDS: Building a Benchmarking Toolkit for Autonomous Robotics

Giulio Fontana, Matteo Matteucci and Domenico G. Sorrenti

Abstract Within computer science, autonomous robotics takes the uneasy role of a discipline where the features of *both* systems (i.e., robots) and their operating environment (i.e., the physical world) conspire to make the application of the experimental scientific method most difficult. This is the reason why much experimental work in robotics is, from the methodological point of view, built on shaky grounds. In particular, scientifically sound benchmarking tools are still largely missing. This chapter starts from RAWSEEDS, a project focused precisely on benchmarking in robotics, to highlight the reasons for these difficulties and to propose strategies for overcoming some of them. The main result of RAWSEEDS is a Benchmarking Toolkit: a readily usable instrument to assess and compare algorithms for SLAM, localization, and mapping. Its most innovative aspects include a set of high-quality, validated, multi-sensor datasets, collected both in indoor and in outdoor locations and complemented by ground truth data, and the explicit definition of a set of quantitative performance metrics for the evaluation of algorithms.

Keywords Robotic datasets · Benchmarking · SLAM · Ground truth

G. Fontana (✉) · M. Matteucci
Dipartimento di Elettronica, Informazione e Bioingegneria, Politecnico di Milano, Milan, Italy
e-mail: giulio.fontana@polimi.it

M. Matteucci
e-mail: matteo.matteucci@polimi.it

D.G. Sorrenti
Dipartimento di Informatica, Sistemistica e Comunicazione, Università degli Studi di Milano-Bicocca, Milan, Italy
e-mail: domenico.sorrenti@unimib.it

F. Amigoni and V. Schiaffonati (eds.), *Methods and Experimental Techniques in Computer Engineering*, PoliMI SpringerBriefs,
DOI: 10.1007/978-3-319-00272-9_4, © The Author(s) 2014

4.1 Introduction

Autonomous robotics is a peculiar discipline. While computing is certainly a key ingredient to it, the added element of physical interaction with the environment changes everything. The presence of large uncertainties (in the outcome of the interactions between robot and environment) and errors (in the perception of the world by the robot) are not the exception, but the rule. Moreover, an autonomous robot determines the course of its actions according to its own assessment of the world: thus, the very behavior of the robot is subject to considerable uncertainty.

This translates into a methodological problem. While the importance of the experimental scientific method to autonomous robotics is as large as it is to other scientific fields associated to computing, the practical difficulties associated to performing accurate experimentation are not. For this reason, methodological soundness often takes a secondary role in robotics papers; while, in the absence of sufficient guarantees of repeatability and/or reproducibility, even the best experimental work tends to take the diminutive role of a mere "proof of concept".

The task of assessing the performance of a robot system (or subsystem) is never trivial; yet, a whole new level of complexity (and cost) must be added to make the results of such assessment usable outside the group which designed the system. For this reason, it is often impossible to quantitatively compare the experimental results obtained with different solutions and/or by different research teams. Over the years, these problems have become increasingly critical to the development of robotics, to the point that today there exists a widespread drive to define reliable benchmarking tools and methodologies [1, 2]. The fact that such tools and methodologies are still largely missing is not due to lack of effort: instead, it is a side effect of strongly heterogeneous experimental conditions, of the weak repeatability of most experimental results, and of the use of subjective and/or insufficiently general performance metrics.

This situation has an even greater impact on industrial research policies. Companies are wary of entering a technological field where marketable applications abound, but heavy investments are needed even to perform preliminary testing of an idea: especially where the lack of established benchmarks also makes technological progess difficult to prove to prospective clients.

The RAWSEEDS project[1] is a benchmarking effort focused on sensor fusion, self-localization, mapping and SLAM in autonomous mobile robots.[2] This chapter describes the development of the RAWSEEDS *Benchmarking Toolkit*, illustrating how the choice of building it on an explicit methodological foundation had an impact on

---

[1] Robotics Advancement through Web-publishing of Sensorial and Elaborated Extensive Data Sets (RAWSEEDS) [4] has been financed by the European Commission, within the 6th Framework Programme.

[2] Sensor fusion is the joint processing of more than one sensor datastreams. Self-localization is the process of finding one's position on a map of the environment. Mapping is the set of operations required to build such a map, usually involving exploration. Finally, Self-Localization And Mapping (SLAM) requires to autonomously build a map of the environment and to keep track of one's location on it.

each stage of the real-world experimental effort of the project. As we will see, the aforementioned impact was (as anticipated in the planning stage) strong in terms of cost, effort and complexity.

The RAWSEEDS Benchmarking Toolkit is composed of:

1. several high-quality multisensor *datasets*, with associated ground truth (GT), gathered by exploring real-world environments (indoor and outdoor) with a mobile robot equipped with a wide set of sensors;
2. a set of *Benchmark Problems* (BPs) built on such datasets, i.e., well-defined problems that also include quantitative criteria to assess their solutions;
3. example solutions to the BPs, called *Benchmark Solutions* (BSs), based on state-of-the-art algorithms and evaluated according to the criteria defined by the associated BPs.

RAWSEEDS is not the only effort towards the definition of benchmarks in SLAM. Other projects with a similar objective exist, the most known of which is the Robotics Data Set Repository (Radish) [3]. However, the RAWSEEDS Benchmarking Toolkit sports novel features, many of which explicitly focus on methodological issues. These include:

- presence of ground truth: each dataset is associated to ground truth data generated independently from the robot sensors;
- strong multisensoriality: the multiple data streams that compose each dataset have been produced at the same time by several sensor systems mounted on the same robot base, thus presenting fully coherent data to subsequent sensor fusion processes;
- data completeness: all data produced by the sensors have been logged in raw form, without performing any data reduction or lossy data compression;
- data synchronization: efforts were devoted to ensure that all data—including those produced by the ground truth collection system—were accurately timestamped according to a single reference clock source;
- data validation and certification: a separate, specific activity was devoted to assess and certify the quality of the datasets and their fitness for their intended purpose;
- wide set of scenarios: datasets have been captured (with the same hardware) indoors and outdoors, under natural and artificial light, in static and dynamic conditions, thus giving to the user the possibility to perform comparisons and verifications;
- explicit evaluation metrics: each Benchmark Problem includes the methodologies needed to evaluate objectively any solution to it, allowing comparison of different solutions independently from the actual choices of implementation and/or representation.

By providing high-quality benchmarking tools to researchers working in SLAM and associated fields, RAWSEEDS empowers them with a way to objectively measure their progress, as well as to compare it to available state-of-the-art algorithms. This is expected to lead to smoother progress (dead-ends are identified earlier), faster recognition by the community of outstanding solutions, and quicker and wider adoption of successful approaches. For what concerns industry, RAWSEEDS aims at providing companies with ready-made tools to evaluate and compare the performance of available algorithms, increasing their confidence towards incorporating them into innovative robotic products. Moreover, the output of the same tools could also be used by companies to demonstrate the quality of their products.

Section 4.2 of this chapter describes the experimental setup, environments and procedures for RAWSEEDS data collection activities. Given the importance of ground truth data within the methodological background of RAWSEEDS, Sect. 4.3 will be dedicated to sketching how they were collected. Section 4.4 focuses on the elements of RAWSEEDS Benchmarking Toolkit, briefly outlining their features. Finally, Sect. 4.5 closes this chapter and outlines the limits of RAWSEEDS.

4.2 Setup and Data Acquisition

As previously said, the RAWSEEDS project focused on methodological issues right from the planning stage. Indeed, many of the methodology-grounded features outlined in Sect. 4.1 had a direct impact both on the design of the robot used to collect sensor data and on the data collection activities. Namely, such features are: strong multisensoriality, data completeness, data synchronization and wide set of scenarios. This section outlines the technological choices made to meet the requirements posed by these features.

4.2.1 Robot Setup

The robot used to acquire RAWSEEDS datasets was composed of two elements: a robot platform and a "sensor frame" module affixed to it, which comprised all the sensors. The robot platform was a custom (i.e., non-commercial) design called *Robocom*, designed for high payload, small dimensions and very good maneuverability. The last two qualities proved crucial for indoor data acquisitions, while the first was necessary due to the significant weight of the sensor frame. During RAWSEEDS data acquisitions, the robot was teleoperated; autonomous navigation would not have added to the quality of the datasets, while requiring additional effort to be set up.

The sensor systems used by RAWSEEDS were chosen to cover a wide range of devices, selected among those which are more frequently used for SLAM. They are:

- Black-and-white binocular and trinocular camera systems.
- Color monocular camera.
- Color omnidirectional camera.
- Four laser range scanners (two high-performance units and two low-cost ones).
- Inertial Measurement Unit.
- Sonar belt comprising 12 ultrasonic transceivers.
- Odometry system based on wheel encoders.

Fig. 4.1 RAWSEEDS data acquisition robot (outdoor setup). SVS is the trinocular camera system, Hokuyo and Sick are the laser range scanners

For outdoor acquisition sessions, the sonar belt was not used (its limited range made it useless), while a GPS receiver and its satellite antenna were added to the robot. Though physically contiguous, the GPS unit was not conceptually part of the sensor suite: it belonged, instead, to the ground truth collection system which will be described in Sect. 4.3. This preserved the independence of the GT collection system from the sensors used for data collection: an important methodological point.

For what concerns calibration, RAWSEEDS provides, as part of the datasets, a full description of the calibration procedures used for each sensor. In the case of cameras, the test images used for calibration are included as well, so that camera calibration can be verified and/or redone by the user. Figure 4.1 shows the RAWSEEDS robot, set up for outdoor data acquisition.

4.2.2 Locations

To avoid resource dispersion, we decided from the beginning of the RAWSEEDS project to perform acquisitions in urban environments only. Within this scope, however, we tried to include a wide set of locations, covering different kinds of environments. RAWSEEDS Benchmarking Toolkit includes datasets recorded in indoor,

Fig. 4.2 A typical view of the Bovisa location (*left*), and an aerial view showing the path followed by the robot during the collection of one of the mixed (i.e., indoor + outdoor) datasets (*right*)

outdoor and mixed (i.e., partially indoor and partially outdoor) urban locations; in natural and artificial light conditions; and in static and dynamic (i.e., with moving objects and people) conditions. Here follows a brief description of the locations featured in RAWSEEDS datasets.

Bovisa. This location is a refurbished factory site, and was used for outdoor and mixed (i.e., outdoor + indoor) datasets. It has a very composite nature, which closely mimics that of a small town or factory area, with buildings separated by roads with sidewalks, complete with parked (and occasionally moving) cars. The dynamic datasets include large quantities of people walking, standing and sitting. The Bovisa location comprises buildings of different kind and style, as well as a wide range of features such as slopes, passages of various widths, external stairs and so on. Robot explorations covered both the outside and the inside of the buildings. Figure 4.2 illustrates the location.

Bicocca. This location was used for indoor datasets only. Corresponding floors of two buildings were involved: an office area in the first building and a library area in the second. The two buildings are interconnected by two glass-walled, roofed bridges, each about 20 m long. The main features of the Bicocca location include: corridors with doors on their sides (some of the doors are deeply recessed within the walls, so corridor boundaries are far from planar); hallways, sporting features such as tables and chairs, columns, staircases and escalators; the two bridges already described; a rather large and architecturally varied library; various kinds of doors and passages. Terrain is very smooth and exclusively horizontal, the only exception being short ramps at one end of the bridges. Figure 4.3 illustrates the location.

Each RAWSEEDS dataset is composed of data collected by the robot while following multiple paths through the environment. Such paths partially overlap and are organized into loops, to trigger the "loop closure" feature of SLAM algorithms. One of the key performance elements for such algorithms is, in fact, the ability to correctly detect that the zone presently explored has been visited before, and to update the map accordingly.

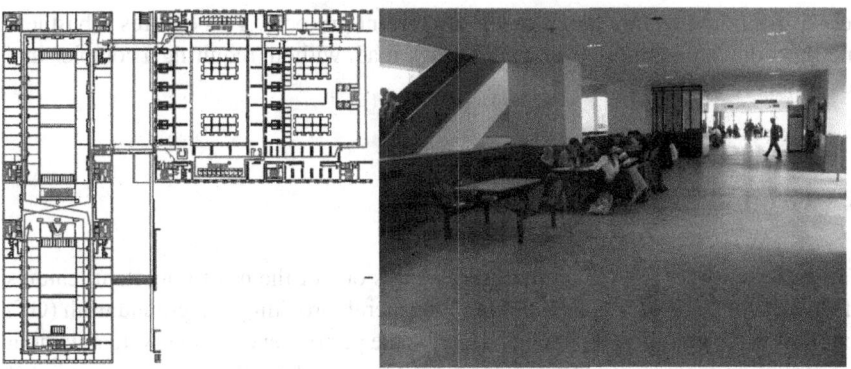

Fig. 4.3 A map of the Bicocca location (*left*), and a typical view (*right*). The map shows the path followed by the RAWSEEDS robot during the collection of one of the datasets

4.2.3 Data Validation and Distribution

All the datasets were subjected to a thorough validation process before making them public, to ensure their validity as the basis for high-quality benchmarking tools. Such process was deemed necessary, notwithstanding the care devoted to the data collection setup and organization, due to the high complexity of the data acquisition hardware and the wide range of environmental conditions. Of all the datasets acquired, those which were considered to be of insufficient quality were discarded. The availability of analyses and certifications of data quality, according to well-specified public criteria, is one of the methodological qualifying points of RAWSEEDS when compared to other available datasets for robotics, which were collected in relation to specific experimental papers.

The data validation process took into consideration several aspects. Specific attention was given to the suitability of the datasets for SLAM algorithms, to ensure the actual usability of the data in the research context to which they were addressed. Specifically, the validation process was aimed at certifying the quality of the following properties of the data: file format; file readability; presence and reliability of timestamp on each data sample; correct time synchronization between data streams; sufficient overlap between successive samples from the same sensor (necessary for tracking of environmental features); density and quality of the data. To ensure that the verification criteria were fully consistent with the intended use of the data, they were checked by actually using them as input for suitable algorithms (feature extraction algorithms for visual data, scan matching algorithms for laser data).

RAWSEEDS Benchmarking Toolkit is freely available trough the project's website http://www.rawseeds.org/, along with the additional documentation needed to make full use of it. Due to the overall mass of downloadable data (around 1.5 TBytes) and to the presence of very large single files (up to 50 GB), the distribution method chosen by RAWSEEDS is very unusual for a scientific dataset: i.e., a peer-to-peer architecture based on the BitTorrent protocol. This ensures robustness and short download times, even under heavy load, without requiring a complex and costly server structure.

4.3 Ground Truth

The availability (or not) of ground truth data is one of the most important features of a dataset intended for benchmarking. In general, providing the ground truth (GT) requires the availability of data that describe the performance of the system without depending on the aspects of the system which are under evaluation. In the case of robots, this means that ground truth must not rely on the same sensor data used by the system for the task under test: therefore separate sensor systems must be deployed. For what concerns errors in the GT, if they are not negligible, they must be known: in this way, their effect on the assessment of system performance can be evaluated and discounted.

Unfortunately, GT collection, especially in autonomous robotics, is difficult and costly, and it was even more in 2006 (when RAWSEEDS collected its datasets).[3] This is the reason why, in robotics, experimental data are seldom accompanied by some form of ground truth. In the case of RAWSEEDS, GT collection accounted for a significant part of the overall effort devoted to setting up and performing data acquisitions.

The ground truth attached to RAWSEEDS datasets takes the forms of maps and (portions of) trajectories. As described by the following of this section, for outdoor datasets GT trajectories were generated with a two-unit RTK GPS system, while for indoor datasets a completely different approach was used (GPS signals are blocked by walls and roofs). For what concerns GT maps, they consist of executive drawings (in the form of CAD files) both for indoor and outdoor environments.

4.3.1 Ground Truth Collection in Outdoor Environments

For outdoor GT collection, RAWSEEDS chose to rely on the established GPS system to gather GT data. Single-receiver GPS systems are insufficiently precise, so a two-unit GPS system (belonging to the class of differential GPS, or DGPS) was required. In our setup, a spatially fixed receiver (*base station*) was used to generate correction signals that are then relayed to the GPS unit to be localized (mounted on the robot and called *rover*) using a dedicated radio link. Conventional DGPS systems estimate location using time shifts between the payloads carried by satellite radio signals: unfortunately, the accuracy afforded by this approach is still insuf-

---

[3] Nowadays the availability of systems for *motion capture* provides a means to precisely acquire the trajectory of a robot. This is still an expensive technology, especially when the capture area is large; however, it is readily available on the market and prices are getting lower.

ficient for an accurate description of robot trajectories. For this reason RAWSEEDS chose a more advanced RTK (Real Time Kinematic) GPS system (several tutorials on RTK GPS systems are available online, such as [5]). These systems use the disalignment between satellite radio carrier signals to estimate timing error, thus yielding greater precision. The price for better precision is paid in terms of increased complexity, difficulty of setup, and price. Under optimal conditions, RTK GPS systems provide centimetre-level positioning errors, which is fully satisfactory for the needs of RAWSEEDS. In the following we will give an account of how frequently these conditions occur in practice.

Our practical experience with the RTK GPS system highlighted some critical issues, many of which are associated to the fact that we were operating in urban environments. In particular, satellite reception was critical, and antenna positioning was a key factor. In theory, a single GPS receiver only requires visibility of 3 GPS satellites to be able to provide a calculated position (called a "GPS fix"). In real-world conditions, due to limitations in the precision of time synchronization between receiver and satellites, visibility of 4 or more satellites is needed. This figure refers to "well-positioned" satellites, i.e., those that have a significant elevation above the horizon, which reduces the number of actually usable satellites. Figure 4.4 shows the RAWSEEDS base station and illustrates RTK GPS performance.

The most damaging effect to reception is due to buildings (and other high obstructions such as trees). As these block GPS signals, obtaining a reliable GPS fix in urban environments can be very difficult. For this reason RAWSEEDS base station was always placed on the top of tall buildings. With this setup, the typical number of (well-positioned) satellites visible from the base station ranged from 6 to 9. Unfortunately, the mobile GPS antenna of the rover was forcibly at ground level, which led

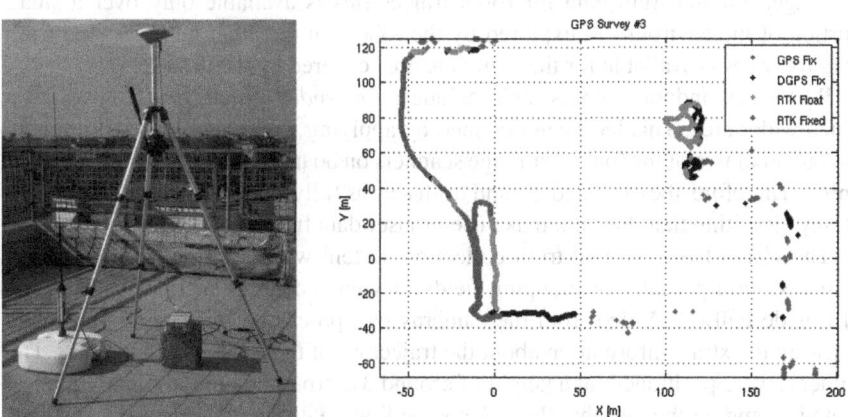

Fig. 4.4 *Left:* RTK GPS base station used by RAWSEEDS to capture ground truth data in outdoor environments. *Right:* a typical example of GPS data acquisition in urban environment. The portions where a best-quality GPS fix is available are the red ones; *light blue* corresponds to a *lower-quality* (but still usable for RAWSEEDS) fix. Building and trees in the immediate vicinity of the robot account for the insufficient quality of GPS data over significant portions of the trajectory

to frequent loss of satellite signals. Situations where the rover received less than the 4 satellites required for the most basic form of GPS fix were very frequent. Things were even more critical for RTK GPS operation, which requires that a minimum of 5 satellites must be available to the base station above the elevation threshold; and, even more critically, the *same* satellites must be visible to the rover. By placing the base station in such a way that all the sky was visible to it, the limiting factor was confined to satellite visibility by the rover.

Even considering its limitations, for RAWSEEDS the RTK GPS system proved to be an effective way to get precise trajectory data over a subset of the robot's complete trajectory. This was, indeed, how it was meant to be used by the project. On the other hand, our experience showed—confirming our expectations—that in outdoor urban environments the usability of GPS systems for robot localization is subject to heavy limitations, and that the use of GPS as a viable alternative method for odometry is questionable.

4.3.2 Ground Truth Collection in Indoor Environments

The collection of ground truth data indoors could not be based on GPS signals, which are blocked. A different approach was chosen: namely, the deployment, in a predefined area, of suitable systems dedicated to the reconstruction of the trajectory of the robot. Two separate systems for indoor GT collection were used in parallel: one based on cameras, the other based on laser range scanners. Their outputs were combined to allow the generation of better-quality GT data [6]. The main limitation of the indoor ground truth collection systems used by RAWSEEDS is their limited coverage. Ground truth data for robot trajectories is available only over a small portion of the environment explored by the robot; on the other hand, ground truth data for maps is available for the complete area covered by the datasets.

RAWSEEDS indoor datasets also include an *extended ground truth*, covering a much wider area. This has been obtained by applying scan matching algorithms to the output of two of the four laser range scanners on board the robot (namely, the Sick units). Therefore, the extended ground truth acts as fully valid GT for the assessment of any algorithm that does not make use of laser data from such sensors.

The vision-based ground truth collection system was based on a network of 5 cameras, with partially overlapping fields of view, covering an L-shaped portion of a wide hallway. Video from the cameras was processed by specially designed software to extract information about the trajectory of the robot (when visible). The cameras were positioned (at a height of around 3 m from the ground) on high poles. Notwithstanding the fact that the poles were fitted with rotation-blocking systems, rotation due to involuntary touches by passers-by, or even to thermal deformations, proved to be an issue: a fact to be considered by anyone planning to use a similar setup.

To allow reconstruction of the trajectory of the robot, the latter was fitted with visual tags and with a rectangular outer "shell" composed of vertical checkered

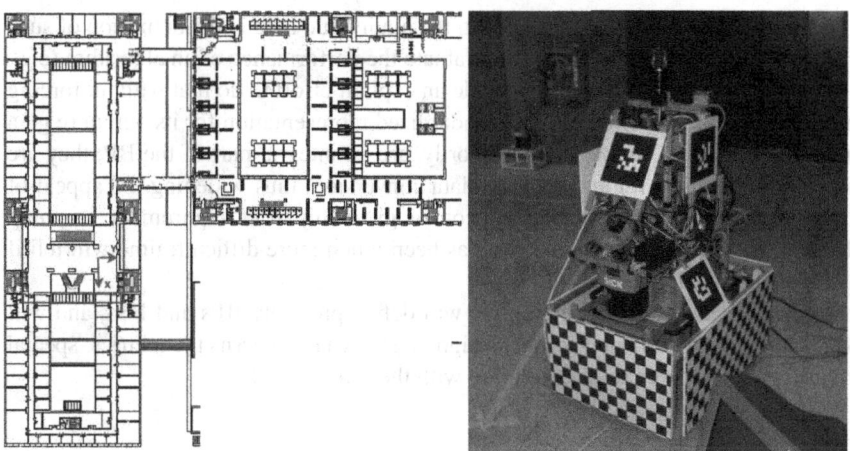

Fig. 4.5 The area of the Bicocca location where ground truth data were collected (*left*) and RAWSEEDS data acquisition robot, set up for indoor data acquisition (*right*). Behind the robot, one of the laser range scanners used by the laser-based GT collection system is visible (a similar device is on board the robot)

boards. This "shell" also had the effect of presenting a more regular shape to the fitting algorithms used by the laser-based GT collection system described below. Figure 4.5 shows Robocom fitted with the tags and "shell" used for indoor GT collection.

The laser-based ground truth collection system was more straightforward. We used a set of four Sick LMS200 laser range scanners positioned in suitable places along the perimeter of the wide hall where GT acquisition was performed. The sensors were carefully aligned so that their perception plane was horizontal, at the optimal height with respect to the robot's "shell" described above. The positions of the laser range scanners were the result of a tradeoff between the conflicting requirements of covering the largest possible area and ensuring that within such area the robot was always perceived by at least two sensors. The output of the laser scanners was processed by scan matching software.

4.4 Benchmark Problems and Benchmark Solutions

The scope of RAWSEEDS is not limited to providing datasets. While these can be immediately used by the community, the Benchmarking Toolkit also includes carefully chosen problems researchers could test their algorithms on, called Benchmark Problems (BPs). Additionally, ready-made solutions to these, based on state-of-the-art algorithms, are provided to act as references: these are called Benchmark Solutions (BSs).

From the methodological viewpoint, the main issue here is the definition of suitable, scientifically sound, *metrics* to evaluate the performance of an algorithm (e.g., a mapping algorithm). RAWSEEDS made an explicit effort to do that without forcing the algorithm developer to adopt a standardized representation for its output (e.g., a map). Such metrics, as we will see shortly, are an integral part of the BP; they are directly applicable without requiring data conversion, thus widening the appeal of RAWSEEDS Benchmark Problems. Of course, given this strict requirement, retaining physical significance for the metrics has been much more difficult; time will tell if such goal has been attained.

In the following of this section we will define precisely BPs and BSs, and will sketch how RAWSEEDS metrics are designed. For what concerns the metrics, special attention will be given to their relation with the ground truth.

4.4.1 Basic Definitions

A **Benchmark Problem (BP)** is the precise description of a type of problem (e.g., "perform SLAM"). The key element of a BP is that it includes not only the definition of a task, but also a set of well-defined performance metrics to assess the output of solutions. By using such metrics, a quantitative evaluation of the quality of the solutions can be done, and different solutions can be quantitatively compared. More precisely, a Benchmark Problem is defined as the union of: (i) the detailed and unambiguous description of a task; (ii) the specifications for a collection of multisensor data, gathered through experimental activity, to be used as the input for the execution of the task; (iii) a rating methodology (i.e., a set of metrics) for the evaluation of the results of the task execution, based on the use of ground truth (GT) data associated to the sensor data. A **Benchmark Problem instance (BP instance)** is obtained by combining a BP with: (iv) one of RAWSEEDS datasets to be used to execute the task specified by the BP.

A solution to a BP instance is called a **Benchmark Solution (BS)** and is defined as the union of: (i) a BP instance; (ii) the detailed description (and, optionally, code) of an algorithm for the solution of the BP instance; (iii) the output of the algorithm when applied to the BP instance; (iv) the values of the metrics specified by the BP, when applied to such output. A BP instance admits any number of different BSs, while a BS is tied to one specific BP instance.

Depending on the BP, the output of a BS can include the map of an environment, the trajectory of the robot, or both. As anticipated, one important point is that the actual representation of both the map and the trajectory are never specified by the BP: this means that any algorithm capable to solve a BP can be easily turned into a RAWSEEDS BS for that BP. The evaluation metrics, as well, are defined so to be computed independently of the actual representation of the output of the BS. In this way, BSs using different representations for maps and trajectories can be compared on performance grounds, which is a very important feature for a benchmarking tool.

4.4.2 Performance Metrics

As previously said, an integral part of each Benchmark Problem instance is a set of representation-independent performance metrics to be applied to its solutions (Benchmark Solutions). The descriptions of RAWSEEDS metrics are *operative*, in the sense that they are intended to be directly applied as an algorithm for their calculation.

The metrics designed by RAWSEEDS to be part of the Benchmark Problems are directly connected to the field of application of the RAWSEEDS Benchmarking Toolkit, i.e., self-localization, mapping, and SLAM. Here follows a short description of them.

Mapping Error. The Mapping Error is intended as a measure of the accuracy of a reconstructed map. It requires that the author of the BS selects a set of environmental features of the map produced by the BS, identifies the corresponding features of the ground truth map, and performs distance calculations within pairs of features belonging to the first set of features and (separately) to the second set of features. A comparison between the results of the distance evaluations performed on the first and second set of features yields the value of the Mapping Error.

Loop Closure Error. This metric is intended to capture the localization accuracy of a SLAM algorithm when it cannot rely on the realigning mechanism called "loop closure", which is triggered when a SLAM algorithm detects that the area where the robot is currently located is one of those already explored by it in the past. This allows to evaluate the quality of the internal map produced by a SLAM algorithm. By eliminating the correction effect of loop closure, error mechanisms that otherwise could be masked are highlighted. The evaluation of the Loop Closure Error requires that the author of the Benchmark Solution disables the loop closure mechanism. Fortunately, considering the way SLAM algorithms are usually implemented this is often a trivial task. Actual computation of the Loop Closure Error is done by comparing the reconstructed and actual (i.e., coming from the ground truth trajectory) pose of the robot at a time instant specified by the BP.

Self-Localization Error. This metric is aimed at evaluating the accuracy of the estimate that a robot produces of its own pose within the environment. Given that such estimate usually refers to a map that is itself reconstructed by the robot, computation of the Self-Localization Error requires that such map is aligned with the ground truth map. This can be done manually, or with any other method, provided that a full description of it is given. When the two maps are aligned, Self-Localization Error is computed by processing the distance errors between the estimated pose of the robot and the corresponding pose from the ground truth trajectory, for each time instant where the latter is available.

Integral Trajectory Error. Like the Self-Localization Error, this is a metric that intends to capture the accuracy of pose reconstruction. Computation is similar, but while Self-Localization Error focuses on instantaneous accuracy, Integral Trajectory Error focuses on the overall distance between the reconstructed and ground truth trajectories (where the latter are available) over the whole path of the robot.

4.5 Conclusion

In a context where the need for benchmarks in robotics is widely perceived but rarely addressed, the RAWSEEDS project is an effort targeted towards fulfilling such need. RAWSEEDS Benchmarking Toolkit is a readily usable instrument to assess and compare algorithms for SLAM, localization, and mapping. Some of its aspects (e.g., strong multisensoriality, focus on ground truth, representation-independent metrics for algorithm assessment) are especially significant from both the methodological and the practical points of view.

Although RAWSEEDS Benchmarking Toolkit has been designed as to be open-ended and extensible, its main limitation lies into its scope. Its components are, in fact, only useful for groups working on the development of algorithms and software which are not involved into the *control* of robots. In fact, the datasets are pre-recorded and therefore not suitable to test algorithms that need to influence the movement of the robot, such as navigation modules. To create a benchmark for this kind of applications, a different approach based on suitable physical *test arenas* [7, 8] seems, at the moment, the only viable alternative.

References

1. Bonsignorio F, Hallam J, del Pobil AP (2007) Good experimental methodology - GEM guidelines. http://www.heronrobots.com/EuronGEMSig/downloads/GemSigGuidelinesBeta.pdf. Accessed July 2013
2. PerMIS (2010) Performance metrics for intelligent systems. http://www.nist.gov/el/isd/ks/permis.cfm, Accessed July 2013
3. Howard A, Roy N (2003) The robotics data set repository (radish). http://www.rawseeds.org/. Accessed July 2013
4. Rawseeds (2006) http://radish.sourceforge.net/. Accessed July 2013
5. Zinas N (2011) GPS network real time kinematic tutorial. Tech. Rep. TEKMON-001, Tekmon Geomatics LLP. http://tekmon.gr/tekmon-research/gps-network-rtk-tutorial/. Accessed July 2013
6. Ceriani S, Fontana G, Giusti A, Marzorati D, Matteucci M, Migliore D, Rizzi D, Sorrenti D, Taddei P (2009) Rawseeds ground truth collection systems for indoor self-localization and mapping. Auton Robot 27(4):353–371
7. Jacoff A, Messina E, Evans J (2002) Experiences in deploying test arenas for autonomous mobile robots. In: Proc 2011 PerMIS, Workshop, pp 87–94
8. Jacoff A, Messina E, Weiss BA, Tadokoro S, Nakagawa Y (2003) Test arenas and performance metrics for urban search and rescue robots. In: Proc 2003 IEEE/RSJ IROS, vol 3, pp 3396–3403

Part III
Computing and Science: Back and Forth

The idea of computing as a science and as an infra-science has represented, in Parts I and II of this volume, a useful guide in the systematization of the different contributions. However, to avoid the risk of a too rigid separation and to show that the two roles are intertwined, this Part III investigates on the mutual relationship between computing and science. The following chapter presents a case, relative to the use of robots to study intelligence, in which the role of support provided by computing raises important and open methodological issues, also in this case with a particular attention to the development of experimental procedures. Even if some theoretical contributions are offered to deal with these questions, no definitive answer is given in order to evidence how many issues at the intersection between computing and science are still open and require, to be properly discussed, other perspectives than the purely scientific-technological ones.

Chapter 5
Biorobotics: A Methodological Primer

Edoardo Datteri

Abstract A first objective of this chapter is to present some interesting roles played by biorobotics in the study of intelligent and adaptive animal behaviour. It will be argued that biorobotic experiments can give rise to different "theoretical outcomes", including evaluation of the plausibility of an hypothesis, formulation of new scientific questions, formulation of new hypotheses, support for broad theses about intelligence and cognition, support for broad regulative principles in the study of intelligence and cognition. These outcomes flow from variants of a common procedure, which will be sketched here. A second objective is to introduce some methodological and epistemological problems raised by biorobotics, which will be analysed in reference to the structure of the common procedure, notably concerning the setting-up and execution of "good" experiments and the formulation of "good" explanations of animal behaviour. Knowing and dealing with these problems is crucial to justifying the idea according to which robotic implementation and experimentation can offer interesting theoretical contributions to the study of intelligence and cognition.

Keywords Biorobotic methodology · Simulations · Robotic modelling · Computational neuroethology

5.1 Introduction

The study of intelligence and cognition has been often supported by robots and computing systems. In the early decades of the twentieth century, the building of robotic systems able to interact dynamically and adaptively with realistic environments contributed to promoting a mechanistic, anti-vitalist approach to the explanation of

E. Datteri (✉)
Dipartimento di Scienze Umane per la Formazione "R. Massa", Università degli Studi di
Milano-Bicocca, Milan, Italy
e-mail: edoardo.datteri@unimib.it

F. Amigoni and V. Schiaffonati (eds.), *Methods and Experimental Techniques*
in Computer Engineering, PoliMI SpringerBriefs,
DOI: 10.1007/978-3-319-00272-9_5, © The Author(s) 2014

animal and human behaviours [1]. More recently, robotic reproductions of animal and insect-like behaviours have provided convincing support for the development of general methodological guidelines for the study of intelligence and cognition and, in some cases, have contributed to formulating broad hypotheses on particular aspects of animal behaviour [2]. Robotic simulations have also been used occasionally to test full-fledged models of particular animal behaviours [3]. Bionic systems, connecting robotic components with living biological tissue, have sometimes been deployed for similar theoretical purposes [4].

The term "biorobotics" is often used to indicate the area of research that makes experimental use of robots as outlined above [5]. Similar roles have been assigned to robots and computer systems in the ages of symbolic Artificial Intelligence and Cybernetics. Biorobotic studies had already been carried out before the advent of computers, one of the most cited examples being the "electric dog" built in 1912 and taken by the mechanistic physiologist Jacques Loeb as a test of his theories on phototropic behaviours in moths [1]. The rapid advancement of robotics and computer technologies is now paving the way for fine-grained simulations of sensory-motor biological mechanisms and realistic reproductions of animal shape. However, this technological progress cannot contribute to solving a number of crucial outstanding *methodological* and *epistemological* issues raised by biorobotics. The objective of this chapter is to outline these issues and to convince the reader of their importance: many of the methodological difficulties introduced here, variously concerning the design and execution of "good" biorobotic studies, are not less serious and urgent than the technological difficulties faced when constructing and programming robotic systems. Addressing these issues may contribute to achieving a deeper understanding of the relationship between computing and science, which is one of the main objectives of this book.

The methodological and epistemological problems discussed here may be grouped into two classes.

- **What makes a "good" biorobotic experiment?** A first group of issues is related to the identification of methodological criteria for the design and execution of "good" biorobotic experiments. This group includes issues concerning the relationship between the biorobot and the theoretical model to be tested, the setting-up of an appropriate experimental scenario and, more generally, justification of the inference of theoretical conclusions about the target *biological* system from *robotic* behaviours.
- **What makes a "good" biorobotic explanation?** Most biorobotic experiments are performed to test *explanations* of intelligent behaviours and cognitive capacities. These studies give then rise to the problem of establishing what constitutes a "good" explanation—or more precisely, the problem of identifying criteria to discriminate between "good" explanations and statements that do not deserve this honorific title.

These issues are introduced and discussed in Sects. 5.3 and 5.4, respectively. Section 5.2 provides an overview of biorobotics methodology, illustrating some interesting roles played by biorobotic experiments in the study of intelligence and cognition, and setting the stage for our methodological discussion.

Before proceeding, it is worth stressing that the issues addressed here are biorobotic variants of issues often addressed by philosophers of science in connection with other domains of scientific inquiry. Philosophy of science is chiefly concerned with the rational justification of scientific research methodologies and with the clarification of basic concepts involved in science. The first group of problems addressed here concern the identification of methodological regulative principles for carrying out "good" biorobotic experiments, "good" experiments being those in which one is *justified* in drawing theoretical conclusions (on animal behaviour) from experimental results (i.e., from the analysis of robot behaviour). The second issue concerns the clarification of the notion of "explanation" in biorobotics. These philosophical issues are of fundamental importance for biorobotics researchers, as the ensuing discussion will show.

5.2 On Various Experimental Roles of Biorobotics

Section 5.2.1 provides an overview of biorobotics methodology, while Sect. 5.2.2 illustrates some interesting roles played by biorobotic experiments in the study of intelligence and cognition.

5.2.1 The Methodological Core

As pointed out in Sect. 5.1, biorobotics can contribute in various ways to the study of intelligence and cognition. These theoretical contributions flow from variants of a common experimental procedure, which is schematically described here with reference to a purely notional biorobotic case study. Figure 5.1 will help follow the various steps of the methodology.

Any given biorobotic study will focus on a particular class L of biological systems (see the top-right box in Fig. 5.1). Suppose L is the class of rats. Typically, biorobotic studies are carried out to explore the mechanisms underlying the manifestation of particular *capacities* CL possessed by members of class L and exhibited by them in particular conditions EL (middle-right box of Fig. 5.1). For example, many biorobotic studies focus on the capacity of rats to orient themselves in experimental mazes. What is a "capacity"? Many philosophers have tried to clarify this notion [6]. Here, consistently with [7], the term will be taken to refer to a (behavioural) *regularity* expressed by a *generalization* statement. For example, rats' capacity to orient themselves in mazes may be expressed as the generalization according to which rats travel from an initial position to a destination point in a maze, making fewer errors on each new trial than on previous trials [8].

Target capacity CL is explained by a hypothetical mechanism description ML (bottom right box of Fig. 5.1). Most mechanism descriptions formulated in the cognitive sciences make reference to a number of interacting *components* within system L, each one playing a distinct role in the exhibition of the target capacity. The notion

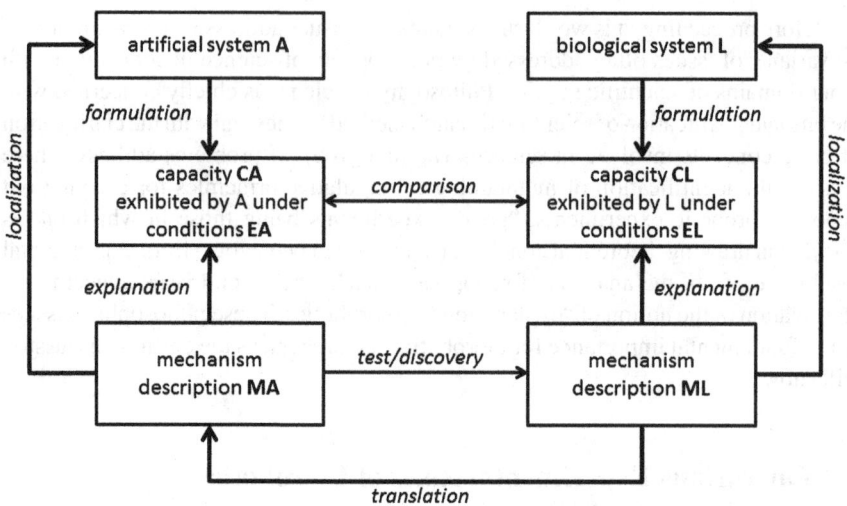

Fig. 5.1 A sketch of biorobotic methodology

of "mechanism" is intrinsically connected with the notion of "regularity": individual components of a mechanism exhibit regular behaviours and interact with each other in regular ways. This is the reason for which cognitive science mechanisms are described by sets of *generalizations*, i.e., of statements expressing regularities in the behaviour of components and their interactions. For example, rat orientation capacities are often explained in terms of a mechanism including a number of sensory and motor organs of the rat and a number of brain areas, notably including the area of the hippocampus containing the so-called *place cells*. The behaviour of these components is assumed to be regular: for example, the firing of each place cell is believed to occur only when the rat is in a particular spatial position, under a number of reasonably well-known boundary conditions [9]. Note that this is a neuroscientific example. Cognitive science generalizations may be couched in non-neuroscientific vocabulary, e.g., vocabulary making reference to mental states such as beliefs, desires and intentions, or to the physical properties of the system, including its morphology [2].

How may we test the hypothesis according to which ML is the mechanism enabling L to exhibit CL? Here is the biorobotic answer: build an artificial (robotic) system A whose behaviour is governed by the mechanism ML under scrutiny, put A under environmental conditions EL, and observe its behaviour (see the left column of Fig. 5.1). This amounts to building a robotic *simulation* of ML and to experimenting on it. If A exhibits capacity CL (i.e., if the robotic simulation reproduces the target biological behaviours), one may be induced to conclude that ML is the mechanism enabling L to exhibit CL. A different result may be taken to support rejection of ML. To be sure, the artificial system A cannot be said to implement ML (which will probably involve *biological* components, such as brain tissues) but a sort of "artificial translation" MA of ML (which will make reference to *artificial* components, such as electronic circuit boards, fulfilling similar functional roles). And the experiments are intended to check, by behavioural comparison, whether the capacity CA exhibited by the artificial system in an experimental environment EA is similar, in a sense to be clarified, to the target capacity CL (see the "comparison" arrow in Fig. 5.1). The "discovery" and the left "explanation" arrows will be discussed later.

5.2.2 Variants and Applications

Experiments with A may play different roles in the study of capacity CL. In particular they can give rise to different "theoretical outcomes", including evaluations of the plausibility of a hypothesis, the formulation of new scientific questions, the formulation of new hypotheses or refinements of previous ones, support for broad theses about intelligence and cognition, and support for broad regulative principles in the study of intelligence and cognition (but the list is not exhaustive). These outcomes flow from variants of the procedure illustrated above, and different outcomes are often obtained at different steps of the same study. Although there may be some overlap, it is worth making these distinctions in order to fully appreciate the value of biorobotics experimentation in the cognitive sciences.

5.2.2.1 Corroboration of Mature Hypotheses

A mechanistic hypothesis ML may be more or less "mature", in the sense of having received more or less support from previous experiments or auxiliary assumptions. In particular, some biorobotic hypotheses are backed by well-corroborated *localization assumptions*, i.e., assumptions stating that some aspects of the mechanism are actually localized somewhere in the target system L. For example, [10] carried out a biorobotic study on a mechanistic hypothesis regarding hippocampal *place cells*, which behaved as prescribed by well-corroborated neuroscientific hypotheses. A large number of features of the mechanistic hypothesis tested in [11] were assumed to be localized in the rat brain and to behave as prescribed by the hypothesis. Mature mechanism descriptions have been tested in biorobotic studies on ant self-localization [12], on cricket phonotaxis [13], on motion detection in locusts [14].

If the hypothesis is already mature in the sense assumed here, why should one perform biorobotic experiments on it? This question points to the unique experimental value of biorobotics (and of simulative methodologies at large). Many conventional experimental methodologies used in the cognitive (neuro)sciences may enable one to evaluate whether aspects of ML are localized in L, e.g., if L has the components mentioned in ML, if these components behave as prescribed by ML, if they are connected as specified by ML. What is missing is the "bridge" between these localization assumptions and the capacity under investigation. Having good reasons to believe that many aspects of the mechanism ML are localized in L is of course crucial to the adequacy of the explanation, but does not amount to claiming that ML *is the mechanism underlying CL.* Some of these aspects could be unrelated to CL. Conversely, some aspects essential to CL could be missing in ML. The hypothesis could well make reference to the "right" components, but connect them in a way that is not functional to CL. Here we may acknowledge one of the distinctive experimental roles of biorobotics (and of computer simulation) in the study of intelligent behaviour: a biorobotic experiment on ML can contribute to assessing whether *these* components, organized in *this* way, produce the behaviour CL of interest. In more general terms, the biorobot crucially contributes to *evaluating the plausibility of a mechanistic hypothesis* ML regarding CL.

Needless to say, maturity comes by degrees: even very mature biorobotic hypotheses may still have aspects that are not backed by well-corroborated localization assumptions. We will discuss this case in Sect. 5.2.2.3.

5.2.2.2 Formulation of New Scientific Questions

Mismatches between CL and CA typically stimulate the formulation of new questions calling for an explanation of the result. For example, the robotic lobster described in [15] consistently failed to match biological performance in a chemiotaxis task under particular conditions. Why? The reason could be sought in biologically unmotivated aspects of the robot (e.g., battery discharge) or in the hypothesized mechanism description. Eventually the authors concluded that robot failures were due to limitations of the proposed hypothesis ML. Biorobotic experiments, in this case, stimulated the *formulation of a new scientific question* which was followed by the formulation of a new mechanistic hypothesis. Note that this question concerns whether the mechanism description ML produces the target behaviour CL or not. In the previous section we noted that biorobotics can play a crucial role in *addressing* questions concerning the "bridge" between mechanisms and their behaviours. Here we add that biorobotics can also play a unique role in *raising* questions of this kind.

Biorobotic experiments may also enable unexpected behaviours of L to be identified, stimulating new why-questions regarding them. The previous example was a case of falsification; now suppose, on the contrary, that A matches L's behaviour to a great extent in a variety of experimental conditions. As discussed so far, this result may be taken to corroborate the hypothesis according to which ML produces CL. Now suppose that A is put in novel internal or environmental conditions EA' producing peculiar behaviours that have never been observed in L (possibly because L's behaviour has never been observed under conditions EA'). Will L display the same peculiar behaviours in EA'? Here, new scientific curiosity regarding L has been raised by experiments with a robotic surrogate. It goes without saying that the same question could have been raised without any robotic experimentation. However, in many cases, manipulation of the robot may be more feasible for practical or ethical reasons than manipulation of the target biological system. And the initial corroboration may support the hypothesis that A and L share the same mechanism, thus increasing the expectation that L will display the peculiar behaviours in EA' and providing strong motivation for addressing the new question.

5.2.2.3 Formulation of New Hypotheses

Let us start from a notional example. Contrary to the cases described so far, suppose that no plausible mechanistic hypothesis on CL is available to fill the "ML" box in Fig. 5.1. This may be due to the fact that previous hypotheses have been discarded, or to the fact that capacity CL has been newly observed. Nonetheless, it is possible to build a robot A whose behaviour CA is similar to CL, by following a conventional, iterative process of robotic design, implementation and testing. If A reproduces the behaviour of interest, one may be induced to "translate" aspects of the mechanism MA implemented in the machine into a new mechanistic hypothesis ML for L, which simultaneously receives initial biorobotic support. This is the "discovery" arrow in Fig. 5.1, the formulation of a new hypothesis being one of the types of discovery taking place in science. In short: if the robot behaves like L, one may be legitimately induced to hypothesize that L produces CL by virtue of the same mechanism implemented in the machine, thus filling the "ML" box. In this case, the role of the biorobotic experiment is to *support the formulation of a new hypothesis regarding CL*. Here is a concrete example. Many robots, built for purposes that are totally unrelated to biological research, produce avoidance and attraction behaviours by virtue of crossed excitatory or inhibitory connections between sensors and motors, as in Braitenberg's vehicles 2a and 2b [16]. These purely robotic implementation successes have stimulated the formulation of a hypothesis about lobster chemiotaxis based on a similar mechanism [15].

It is worth noting that this example and those discussed in the previous sections share important aspects of the methodological procedure described in Sect. 5.2.1: in both cases, behavioural comparisons between A and L provide elements for reflection on the similarity between MA and ML. The two cases differ with regard to the maturity of the hypothesis. A relatively mature hypothesis was available before robotic implementation in the examples discussed in Sect. 5.2.2.1 while in case just outlined no previous hypothesis is available: it is newly formulated via translation from MA.

Processes of translation from robotic mechanism descriptions to biological hypotheses often take place in biorobotics, even when a biological hypothesis ML is available. MA may have features that are not reflected in ML—as we will discuss in Sect. 5.3, all biorobots include aspects that are not mentioned in the biological hypothesis under scrutiny. In particular, MA may include *components* that are necessary for producing the desired behaviour even though they are not mentioned in ML (e.g., because scientific theorizing on CL is still in its infancy). In this case, success on the part of A in replicating the behaviours of L may induce one to include those components in ML as well. For example, the robot described in [10] included so-called artificial *goal cells* which were necessary to memorize goal locations, and therefore to build a system able to fully replicate the maze navigation capabilities observed in rats. In the experiments the robot displayed goal-seeking abilities, and this was taken to support the hypothesis that something functionally equivalent to goal cells could be found somewhere in the rat brain: the experiment supported the *formulation of a new localization assumption* to the effect that goal cells can be found in the rat brain.

In other cases a *mechanistic hypothesis*, and not simply a localization assumption, is obtained by translation from MA. In the study on cricket phonotaxis described in [13], the robot was found to be affected by environmental conditions that do not affect "real" cricket behaviour: the grass on which both systems were placed slowed down the robot. This was not only due to the particular shape of the robot, but also— as argued by the authors—to the fact that it moved without any feedback-based control mechanism able to correct deviations due to the grass. An obvious solution to this robotic problem would be to provide the system with a feedback-based control mechanism of that sort. This suggestion gives naturally rise to a new mechanistic hypothesis about crickets: they are likely to have some form of feedback-based mechanism to overcome the friction caused by the grass and move efficiently on it.

Finally, *refinements of previous hypotheses* on L are very often obtained by translation from MA. This typically happens when the initial hypothesis under scrutiny ML is formulated in vague terms, e.g., when it includes unfixed parameters that must be fixed in order to obtain a working system [17]. For example, the mechanism description ML tested in the aforementioned study on lobster chemiotaxis [15] included no prescription regarding the distance between the two chemical sensors to be put at each side of the robot. Needless to say, this parameter must be fixed in order to build the robot—the sensors must be put at *some* distance! Let us call ML' the mechanism description ML with distance value fixed. As in the previous examples, ML' is obtained by "translation" from MA: the exact distance value is specified in a description MA of the mechanism implemented in the robot. ML and ML' are clearly different in at least one respect—not a trivial one indeed, as intra-sensor distance *may* actually matter for a robot whose behaviour is dependent on the difference between the stimuli perceived at each of two sensors. And one may legitimately claim that the hypothesis actually tested in the study is the *latter* one: strictly speaking the robot must be considered as a simulation of the *refined* mechanism description ML' rather than of the vaguer hypothesis ML. This example will be discussed again in Sect. 5.3 in connection with the issue of evaluating simulation accuracy.

5.2.2.4 Constructive Proof of Mechanicism

We have argued that biorobotic experiments can contribute to formulating novel mechanistic hypotheses about the behaviour of living systems and their cognitive capacities. In some cases, especially in the early decades of the twentieth century, this has amounted to supporting *mechanicism*, understood as an epistemological orientation towards explaining events by identifying the mechanism producing them. *Vitalist* philosophers and physiologists, including Henri Bergson and Hans Driesch, believed that many aspects of intelligence and cognition, including learning, could be explained only by appeal to *non-mechanical* vital forces. Let us call CL a general description of a particular cognitive or behavioural capacity. In many cases, the construction of a robot exhibiting a behaviour (CA) substantially similar to CL has provided support for the thesis that CL *may be explained mechanistically*: insofar as at least *one* mechanistic explanation of CL had been found to exist, namely the mechanism implemented in the machine A (in our terms, the mechanism ML obtained by translation from MA). Many such cases are discussed in [1], including Ashby's homeostat, Hull's 'psychic machine', Grey Walter's tortoises and, in more recent times, Braitenberg vehicles (see also [18]).

5.2.2.5 Guidelines for Explaining Intelligence

The so-called 'embodied approach' in Cognitive Science and Artificial Intelligence is based on a number of broad theses concerning the nature of intelligent behaviour and the appropriate way to explain it. The most basic of these propositions will by now seem obvious to most of us: the behaviour of a living system is not only determined by the control mechanism implemented in it but also by its interaction with the external world—and a "simple" control mechanism can produce "complex" behaviours due to the "complexity" of the environment, as stressed in [16, 19], and others. This claim gives rise to a variety of broad theses regarding the most appropriate way to build efficient robots and explain intelligent behaviour. As far as explanation is concerned, it is stressed that in explaining the behaviour of a system, particular attention should be paid to its shape and to the features of its *ecological niche*. This may help to avoid what Braitenberg believed to be a bias typical of (cognitive) scientists, that is to say, the tendency to explain "complex" behaviours by appeal to "complex" mechanisms.

This *guideline for explaining intelligence and cognition* has been supported by the implementation of robots able to exploit their shape and physical dynamics, rather than sophisticated control mechanisms, to generate apparently "complex" behaviours. Cases in point are the Swiss and Stumpy robots, developed at the Artificial Intelligence Laboratory of the University of Zurich [2]. The robot-based methodology supporting this guideline for explanation fits well with the procedure described in Sect. 5.2.1, and the case we are discussing is substantially similar to those discussed in Sect. 5.2.2.3. Robot A generates behaviours that are very similar to those observed in a broad class L of living systems. The mechanism used, MA, exploits particular features of A's shape and environment. Similarly to the case discussed in Sect. 5.2.2.3, this result may be taken to support the broad claim that L's behaviour may be explained by a sort of "biological translation" of MA, i.e., that L's behaviour can be explained by appeal to particular features of L's shape and environment. As a result, the robot may be regarded as a positive implementation of the guideline for explaining intelligence outlined above, according to which particular attention should be paid to the shape of a system and to the features of its ecological niche in explaining its behaviour.

5.3 What Makes a Good Biorobotic Experiment?

We have discussed various ways in which biorobotics can contribute to the study of intelligence and cognition. The procedures leading to these results share the common methodological structure illustrated in Sect. 5.2.1. And they also share a number of epistemological and methodological problems affecting that methodological structure. Figure 5.1 may help to identify these problems, some of which are related to the design and execution of a "good" biorobotic experiment.

5.3.1 Experimental Comparisons Between CA and CL

Webb [20] has convincingly argued that every biorobotic inquiry regarding intelligence and cognition *in living systems* must be based on some kind of comparison between biological and robotic behaviours. No interesting insight into animal behaviour can be obtained by reasoning solely about robot behaviours, contrary to what has been suggested by proponents of the so-called "animat" approach within robotics. In "good" biorobotic experiments, one draws theoretical conclusions about ML from the result of experimental comparisons between CA and CL. This immediately gives rise to a methodological justification problem: how should these comparisons be carried out for their results to play a legitimate role in the testing and discovery of ML?

In particular, what aspects of the two behaviours should be considered in the comparison? In the aforementioned study on lobster chemiotaxis, for example, the authors focused on the success rates of robotic and "biological" lobsters in reaching the destination site, irrespectively of the trajectories followed by the two systems. Finer-grained comparisons of robotic and human elbow trajectories were made in the biorobotic study on forearm posture maintenance described in [21]. Clearly, the outcome of the comparison between CA and CL (and, consequently, the outcome of the whole biorobotic study) crucially depends on, amongst other factors, the particular aspects being compared: the robot may match L's rate of success in reaching the destination by following completely different trajectories. How may "good" matching criteria be chosen?

This is by no means an easy methodological question, and here we can only provide some prompts for further discussion. In principle, the range of possible matching criteria is very wide in any biorobotic study. However, it is reasonable to claim that what constitutes the "right" criteria *depends on the scientific question* addressed in the study. If one aims to explain why lobsters' rate of success in reaching the source of a chemical stream is so high, then one should look at the rate of success of the robotic simulation. If, instead, one aims to explain why lobsters generate certain specific trajectories as opposed to others, then one should compare robot and animal trajectories. Conversely, comparisons between rates of success will legitimately enable one, at most, to theorize on the animal success rate and not, in principle, on the trajectories generated by the living system.

A related crucial methodological question concerns the setting-up of the experimental setting EA in which to observe robot behaviours. Ideally, in most cases, robots can operate in environmental conditions that are very similar or identical to the animal's ecological niche: an example is the robot used to study the localization abilities of the Cataglyphis desert ant, which is used in the Sahara desert [12]. However, in many studies, robot and biological behaviours are observed in quite different environments [10]. If EA and EL are substantially different, is one justified in taking A's behaviours as empirical evidence in reasoning about the mechanism producing CL in EL? In principle, the analysis of A's behaviour in EA could enable one, at most, to theorize on the behaviour produced by L in EA. However, on closer scrutiny, this seems to be too strong a position: one may reasonably claim that some degree of resemblance between EA and EL licenses *some kind* of theoretical conclusion regarding L's behaviour in EL. This methodological justification problem calls for the identification of regulative principles governing the set up of "good" experimental environments in biorobotics. To address this problem it is worth stressing that why-questions investigated by the cognitive sciences typically do not concern animal behaviours observed in their ecological niche, but rather behaviours observed in specifically tailored and controlled experimental settings. Robots can go wild [22], but biorobotic experiments are often designed to test hypotheses on animal behaviours observed in laboratory settings [23].

5.3.2 Simulation Accuracy

Not *every* robot can contribute to testing a mechanistic hypothesis ML. Needless to say, a commercial robotic vacuum cleaner can hardly provide empirical evidence to test a hypothesis on human posture maintenance. There must be a close relationship between ML, MA and A in order to make legitimate use of A in the testing of ML. In particular, it has been often claimed that the robot itself should be a *good simulation of the hypothesis ML under scrutiny*. Otherwise, it is not clear why robotic behaviours should be taken as empirical evidence in reasoning about ML. But what is the nature of this close relationship? What makes a good robotic simulation of a mechanistic hypothesis? In other words, what kind of criteria should be used to check if A is a good simulation of ML?

According to a plausible interpretation of the term "simulation", A is a good simulation of ML if A works *as prescribed by ML* or, equivalently, if A implements the mechanism described in ML. However, on closer scrutiny, this condition seems hard to attain. First, as already noted in Sect. 2.2.3, cognitive science mechanistic hypotheses are often vaguely and qualitatively specified; any robotic implementation of them may be regarded, at least in principle, as the implementation of a *fully specified* version of the initial hypothesis. Second, all robots will include components that are not mentioned in ML: animals do not need DC batteries. Third, the process of robotic implementation often involves approximations and adjustments with respect to the initial mechanism. For these and other reasons (more extensively discussed in [17] and [3]), it seems unlikely that the mechanism MA actually implemented in the machine will be *exactly the same as* the mechanism ML under scrutiny. In this case, are there rational grounds for justifying the use of A to reason about ML?

This question is still open and is not easily answered. Indeed, in most biorobotic studies, it is simply claimed that the robot "is closely based on", "implements accurately", "simulates in detail" the target hypothesis, but these claims are not well clarified and justified. And in many cases, closer scrutiny will identify non-trivial discrepancies between the hypothesis and the implemented mechanism. The problem of defining criteria for evaluating simulation accuracy has been occasionally addressed in the methodological literature (e.g., [24]) but a satisfactory solution is still lacking. Without purporting to solve the issue here, let us briefly propose a means of viewing this problem from another, possibly more fruitful, perspective.

We have reasoned about the possibility of building a robot A that behaves *exactly like* the hypothesis ML under scrutiny. And we have pointed out that every biorobot A possesses some features that are not specified by the biological hypothesis and are therefore determined on the basis of other criteria. However, we should avoid jumping too quickly from the existence of these differences to the conclusion that A is a bad experimental tool for testing ML. This point can be aptly illustrated by reference to the experiment on lobsters chemiotaxis described in [15]. As outlined above, ML contained no prescription regarding the distance between the chemical sensors: it was a vague hypothesis at least in this respect. This ambiguity had to be addressed in order to build A, and the fixed intra-sensor distance value was specified in a description MA* of A (the reason for the asterisk will become clear later). So, there was at least one difference between MA* and ML. However, the experiments showed that distance value was totally *irrelevant* to the robot's ability to replicate the behaviour of interest. Indeed, the authors conducted several experiments with different intra-sensor distances, finding that the level of behavioural match between biological and artificial behaviours did not change. Therefore the difference between MA* and ML was irrelevant with respect to the outcomes of the behavioural comparisons between the two systems—the particular intra-sensor distance chosen by the authors *did not make the difference with respect to whether A replicated L's behaviours or not*, in the sense that different intra-sensor distances did not give rise to different behaviours (i.e., they did not increase the robot's success rate). According to many plausible accounts of what it is to explain something (see, e.g., [25]), we therefore would not mention intra-sensor distance in a mechanistic explanation of A's ability to reproduce L's behaviours. For similar reasons, the fact that A's external structure is made of polyethylene instead of polypropylene would not be mentioned in a mechanistic explanation of A's behaviours, provided that this choice does not make any relevant behavioural difference. Let us call MA this mechanistic explanation, which is silent on intra-sensor distance. Now, it seems reasonable to claim that MA is superior to MA* in describing *the* mechanism producing CA: MA and not MA* includes a specification of the aspects that are actually *relevant* to A's behaviour. So, why worry about the difference between ML and MA*? It is reasonable to claim that what is important is whether there are differences between ML and MA, i.e., whether there are differences between ML *and the mechanisms that are really relevant to the manifestation of CA*.

The conclusion of this reflection is that not *every* difference between the hypothesis under investigation and the robot is relevant to whether the latter is a good tool to reason about the former. What matters is whether the biological hypothesis is similar, in a sense to be specified, to the mechanism MA which has actually governed the robot in the experiments; and, for the reasons discussed here, not every peculiarity of the robot needs to be specified in MA. In this perspective, evaluation of simulation accuracy crucially requires the formulation of good explanations of *robotic* behaviours. This claim, which is left to the reader as an insight for further analysis, introduces us to another fundamental methodological issue concerning biorobotics.

5.4 What Makes a Good Biorobotic Explanation?

Biorobotic studies are typically aimed at formulating *good explanations* of biological behaviours (see the arrow labelled "explanation" connecting ML and CL in Fig. 5.1). However, no clear and precise criteria are available to distinguish "good explanations" of a given event or regularity from statements that do not deserve this title. Many substantially different mechanistic hypotheses may be formulated to explain capacity CL—how should we choose among them?

A bioroboticist may suggest using a biorobotic test: if a robotic simulation of ML reproduces the target behaviour, then accept ML as a good basis for explaining CL. However, this is only part of the story. Simulation success may corroborate ML but it seems to be *insufficient* to conclusively claim explanation adequacy. The feedback-based hypothesis ML about lobster chemiotaxis described above prescribes that each chemical sensor is positively connected to the motor organs located on the opposite side of the system. Suppose that a robotic simulation of this hypothesis performs efficient chemiotaxis. This result may induce one to corroborate ML, but will not dispel all doubts about its explanatory adequacy. First, ML is very "simple": no mention is made of the mechanisms actually connecting sensors with motor organs (there is surely more than a pair of excitatory neurons) or of the gait control mechanisms. Second, ML is very "idealized": the hypothesis only makes sense if we assume that no external or internal perturbation will affect lobster behaviour (no mechanism is included to resist water turbulence, avoid predators, choose between competing internal motivations, and so on). One may claim that these are not serious objections and that ML is a good basis for explanation even though it is simple and idealized. However, this claim needs justification—and it cannot be adequately justified without appealing to some notion of what makes a good explanation!

This is clearly a crucial and urgent methodological issue for anyone who aims to *explain* intelligent animal behaviours. And it is still an open issue, similarly to those discussed in the previous sections. A thorough discussion is beyond the scope of this chapter. Our aim is merely to provide some insights for discussion, by providing a brief overview of the two main positions put forward in the philosophical literature regarding this problem.

Let CL refer to the statement expressing what is to be explained, for example, "rats travel from an initial position to a destination point in a maze, making fewer errors on each new trial than on previous trials". And let "Exp" refer to the statements constituting the proposed explanation, for example, statements describing a neural mechanism *plus* other statements specifying some initial or boundary conditions. According to the so-called *ontic* view of scientific explanation, Exp is a good explanation of CL if and only if a particular relationship—i.e., a *causal relationship*—holds between what is described by Exp and what is described by CL. In short, a good explanation of something describes its causes. This is a very plausible and commonsensical position ("explaining" is often used as synonymous of "finding the causes" in the everyday language). However, it is seriously affected by the difficulty encountered in defining the notion of a "cause"—or more precisely, by the lack of criteria for distinguishing causal relationships from non-causal generalizations [26].

An alternative position on the nature of scientific explanation is the so-called *epistemic* view. The idea is that Exp is a good explanation of CL if and only if a particular relationship holds between *knowledge of* what is described by Exp and *knowledge of* what is described by CL (the italics mark the difference with respect to the ontic thesis). In particular, many supporters of the epistemic approach claim that, in a good explanation, knowledge of what is described by Exp should have allowed one, if taken into account in time, to *predict* what is described by CL (prediction being a sort of "knowing in advance"). This is a plausible idea, at least in general terms. The feeling of having received a good explanation of an event is very often accompanied by the feeling that, had we known the explanation in time, we could have predicted that event. This view is not affected by the many problems arising from the notion of "causation", simply because it does not include that notion. However, particular epistemic models of scientific explanation—notably the so-called deductive-nomological model [27]—have been widely criticized in the philosophical literature [26].

As far as *biorobotic* explanations are concerned, there are good reasons to claim that they presuppose an epistemic account of scientific explanation according to which the ability to predict the behaviour of interest is at least *essential* (though not sufficient) for a good explanation. The main reason is that, in this methodology, robotic systems are used to identify implications—which in many cases consist of *predictions*—of the explanatory hypothesis ML under scrutiny. This claim needs further clarification and justification, which is beyond the scope of this brief introduction to the philosophical problem of scientific explanation. As previously noted, the problem is still open: careful analyses of "good" and "non-good" scientific explanations, possibly drawing on the biorobotic literature, are needed to solve it.

5.5 Summary and Conclusion

A first objective of this chapter was to present some interesting roles played by biorobotics in the study of intelligent and adaptive animal behaviour. We have claimed that biorobotic experiments can give rise to different "theoretical outcomes", including evaluation of the plausibility of an hypothesis, formulation of new scientific questions, formulation of new hypotheses, support for broad theses about intelligence and cognition, support for broad regulative principles in the study of intelligence and cognition. We have shown that these outcomes flow from variants of a common procedure. A second objective was to introduce some methodological and epistemological problems raised by biorobotics, which we have analysed in reference to the structure of the common procedure, notably concerning the setting-up and execution of "good" experiments and the formulation of "good" explanations of animal behaviour. Knowing and dealing with these problems is crucial to justifying the idea according to which robotic implementation and experimentation can offer interesting theoretical contributions to the study of intelligence and cognition, and may contribute to achieving a deeper understanding of the relationship between computing and science, which is one of the main objectives of this book.

References

1. Cordeschi R (2002) The discovery of the artificial: behavior, mind and machines before and beyond cybernetics. Kluwer Academic Publishers, Dordrecht
2. Pfeifer R, Bongard J (2007) How the body shapes the way we think: a new view of intelligence. The MIT Press, Cambridge (MA)
3. Datteri E, Tamburrini G (2007) Biorobotic experiments for the discovery of biological mechanisms. Phil Sci 74(3):409–430
4. Datteri E (2009) Simulation experiments in bionics: a regulative methodological perspective. Biol Phil 24(3):301–324
5. Webb B (2001) Can robots make good models of biological behaviour? Behav Brain Sci 24:1033–1050
6. Cartwright N (1994) Nature's capacities and their measurement. Oxford University Press, Oxford
7. Cummins R (1985) The nature of psychological explanation. The MIT Press, Cambridge (MA)
8. Tolman E (1948) Cognitive maps in rats and men. Psychol Rev 55(4):189–208
9. Moser E, Kropff E, Moser M (2008) Place cells, grid cells, and the brain's spatial representation system. Annu Rev Neurosci 31:69–89
10. Burgess N, Jackson A, Hartley T, O'Keefe J (2000) Predictions derived from modelling the hippocampal role in navigation. Biol Cybern 83(3):301–312
11. Krichmar J, Seth A, Nitz D, Fleischer J, Edelman G (2005) Spatial navigation and causal analysis in a brain-based device modeling cortical-ippocampal interactions. Neuroinformatics 3(3):197–222
12. Lambrinos D, Möller R, Labhart T, Pfeifer R, Wehner R (2000) A mobile robot employing insect strategies for navigation. Robot Auton Syst 30:39–64
13. Reeve R, Webb B, Horchler A, Indiveri G, Quinn R (2005) New technologies for testing a model of cricket phonotaxis on an outdoor robot. Robot Auton Syst 51(1):41–54

14. Blanchard M, Rind F, Verschure P (2000) Collision avoidance using a model of the locust LGMD neuron. Robot Auton Syst 30(1–2):17–38
15. Grasso F, Consi T, Mountain D, Atema J (2000) Biomimetic robot lobster performs chemoorientation in turbulence using a pair of spatially separated sensors: progress and challenges. Robot Auton Syst 30(1–2):115–131
16. Braitenberg V (1986) Vehicles: experiments in synthetic psychology. The MIT Press, Cambridge (MA)
17. Tamburrini G, Datteri E (2005) Machine experiments and theoretical modelling: from cybernetic methodology to neuro-robotics. Mind Mach 15(3–4):335–358
18. Schlimm D (2009) Learning from the existence of models: on psychic machines, tortoises, and computer simulations. Synthese 169:521–538
19. Simon H (1969) The sciences of the artificial. The MIT Press, Cambridge (MA)
20. Webb B (2009) Animals versus animats: or why not model the real iguana? Adapt Behav 17(4):269–286
21. Chou P, Hannaford B (1997) Study of human forearm posture maintenance with a physiologically based robotic arm and spinal level neural controller. Biol Cybern 76(4):285–298
22. Knight J (2005) Animal behaviour: when robots go wild. Nature 434:954–955
23. Datteri E, Laudisa F (2012) Model testing, prediction and experimental protocols in neuroscience: a case study. Stud Hist Phil Biol Biomed Sci 43(3):602–610
24. Webb B (2006) Validating biorobotic models. J Neural Eng 3(3):R25–R35
25. Woodward J (2003) Making things happen: a theory of causal explanation. Oxford University Press, New York
26. Psillos S (2002) Causation and explanation. Acumen - McGill Queens University Press, Georgetown, Canada
27. Hempel C, Oppenheim P (1948) Studies in the logic of explanation. Phil Sci 15(2):135–175

Concluding Remarks

While the multiple interconnections between computing and science have been elsewhere recognized and observed under several different perspectives, in this volume we have chosen to focus on the methodological one, and in particular on the experimental method. Not only computing acts as an increasingly important, and in many cases necessary, tool for supporting scientific endeavors, but also the scientific nature of computing itself has been debated since the origin of the discipline. It is a matter of fact that computing presents a peculiar methodological approach that shows both similarities and dissimilarities with the traditional experimental scientific method, and that it is definitely worth further investigating.

Science and computing have progressively interacted in the last years and many significant results have been achieved. In this volume we have rather emphasized some of the many issues that are still open. Accordingly, we have not provided any definitive answer, but we have tried to raise questions and to open the way for future investigations, acknowledging the importance of a real interdisciplinary approach for such research.

Finally, stemming this volume from an educational experience (a course for graduate students in the Ph.D. program in Information Engineering at the Politecnico di Milano), we wish to highlight the importance of discussing these challenging topics not only from a research point of view, but also from an education perspective. Future computer scientists and engineers should grow with an increasing awareness of their discipline at the intersection of several fields, as a science itself—even of a very peculiar type—and as a infra-science at the service of other sciences.

F. Amigoni and V. Schiaffonati (eds.), *Methods and Experimental Techniques in Computer Engineering*, PoliMI SpringerBriefs, DOI: 10.1007/978-3-319-00272-9, © The Author(s) 2014